Series on Analysis, Applications and Computation – Vol. 8

ISAAC

The Linearised Dam-Break Problem

Series on Analysis, Applications and Computation

Series on Analysis, Applications and Computation – Vol. 8

ISAAC

The Linearised Dam-Break Problem

∘ D. J. Needham
University of Birmingham, UK

∘ S. McGovern
University of Birmingham, UK

∘ J. A. Leach
University of Birmingham, UK

World Scientific

NEW JERSEY · LONDON · SINGAPORE · BEIJING · SHANGHAI · HONG KONG · TAIPEI · CHENNAI · TOKYO

Published by

World Scientific Publishing Co. Pte. Ltd.

5 Toh Tuck Link, Singapore 596224

USA office: 27 Warren Street, Suite 401-402, Hackensack, NJ 07601

UK office: 57 Shelton Street, Covent Garden, London WC2H 9HE

Library of Congress Cataloging-in-Publication Data

Names: Needham, D. J. (David J.), author. | McGovern, S. (Stewart), author. |
    Leach, J. A. (John Andrew), 1965–   author.
Title: The linearised dam-break problem / by D.J. Needham (University of Birmingham, UK),
    S. McGovern (University of Birmingham, UK), J.A. Leach (University of Birmingham, UK).
Description: New Jersey : World Scientific, 2017. | Series: Series on analysis, applications,
    and computation ; 8 | Includes bibliographical references.
Identifiers: LCCN 2017028442 | ISBN 9789813223875 (hardcover : alk. paper) |
    ISBN 9789813223882 (ebook) | ISBN 9789813223899 (ebook other)
Subjects: LCSH: Fluid dynamics--Mathematical models. | Dam failures--Mathematical models. |
    Hydrodynamics. | Momentum (Mechanics)
Classification: LCC QA911 .N394 2017 | DDC 532/.05--dc23
LC record available at https://lccn.loc.gov/2017028442

British Library Cataloguing-in-Publication Data
A catalogue record for this book is available from the British Library.

For any available supplementary material, please visit
https://www.worldscienti ic.com/worldscibooks/10.1142/10543#t=suppl

Printed in Singapore

Contents

Chapter 1

Introduction

A dam is a barrier that separates two horizontal layers of fluid from each other. Generally, a dam is a man made structure that is used to hold a body of water in place, forming a reservoir or lake, which can then be used to support local communities. Dams can be used in order to supply water to local communities, or act as a flood defence. They can also be used to supply water for farmland and to allow hydro-power electricity, for further examples, see [via DIALOG. (2015)]. The failure of a dam will generally have a catastrophic impact on the surrounding environment. As a consequence the study of the hydrodynamical effects of dam failure is of principal significance. The problem of a dam failure is modelled mathematically as the classical dam-break problem. It is developed to model the situation where a mass of fluid, lying on an impermeable bed, is held in place by a solid boundary (the dam) so that there is a change in the fluid surface elevation across the dam site. The dam-breaks at an initial time and the aim is to determine the behaviour of the subsequent hydrodynamic flow; that is, to determine the fluid velocity field and, in particular, the free surface behaviour. The dam-break flow is formulated mathematically by the conservation of mass and momentum equations for the fluid flow, together with appropriate boundary and initial conditions. This leads to a free boundary nonlinear evolution problem; it is therefore convenient to make some reasonable approximations to reduce the difficulty in analysing the dam-break problem. As this is a water wave problem, viscous effects are ignored and the fluid is taken to be incompressible. It is also assumed that the dam disappears instantly at the initial time. These assumptions are standard in dam-break problems, see for example [Ungarish et al. (2014)], [Yilmaz et al. (2013)], [Korobkin and Yilmaz (2009)], [Ancey et al. (2008)], [Fernandez-Feria (2006)], [Hunt (1983)], [Ostapenko (2003)].

The majority of work on the dam-break problem employs from the outset the shallow water approximation, see for example [Zhou et al. (2002)], [Zoppou and

1

Roberts (2000)], [Ungarish *et al.* (2014)], [Brufau and Garcia-Navarro (2000)], [Ancey *et al.* (2008)], [Ahmad *et al.* (2013)], [Hunt (1983)], [Ostapenko (2003)], [Ostapenko (2007)], [Hogg (2006)], [Goater and Hogg (2011)]. This assumes that the fluid pressure is hydrostatic on the basis that the vertical acceleration of the fluid has a negligible effect on the pressure. By integrating the equation for conservation of mass and horizontal momentum over the vertical fluid layer thickness and then applying the hydrostatic pressure assumption, the shallow water equations are obtained. These approximate the horizontal fluid velocity field and the free surface. A detailed derivation is given in [Stoker (1957)] (p. 22). Furthermore, assuming that the horizontal component of fluid velocity and the surface elevation are small, the shallow water equations can be linearised and combined to give the linear wave equation as the governing partial differential equation for the free surface evolution (see [Stoker (1957)], p. 24).

In the classical text on the theory of water waves by Stoker, [Stoker (1957)] (Chap. 10 p. 308), the shallow water equations are used to model the case of a two-dimensional dam-break problem, where a mass of fluid, initially at rest, sits on a flat, horizontal, impermeable bed, and is held in place by a piston so that there is a constant fluid depth (which extends infinitely far in the horizontal direction) on one side of the piston, and there is zero fluid depth on the other side of the piston. The piston is pulled away from the fluid so that a depression wave is initiated. In particular, the case where the piston is pulled away impulsively was considered in detail. This corresponds to a dam-break problem where the dam disappears instantly. The problem was solved by the method of characteristics so that the surface elevation and horizontal component of fluid velocity were determined. The shallow water theory developed in [Stoker (1957)] has since been used in many other situations, for example a dam-break problem on a bed with a step, [Ostapenko (2003)], for arbitrary slopes of the bed, [Fernandez-Feria (2006)], dam-breaks over a bed with a drop, [Ostapenko (2003)], and for the modified shallow water equations over a dry bed, [Ostapenko (2007)].

Numerical approaches have also been adopted to gain insight into dam-break flows. In [Sopta *et al.* (2007)], the shallow water equations were approximated numerically to predict the potential damage caused by a dam-break on the village Tribalj in Croatia, and their results were used in the urban planning of the flood risk areas. Many other numerical approaches in the shallow water theory have been developed, see for example [Hu and Sueyoshi (2010)], [Zoppou and Roberts (2000)], [Zhou *et al.* (2002)], [Xing and Shu (2005)], [Wubs (1988)].

Closely related to the work presented in this monograph is the asymptotic analysis developed in [Leach and Needham (2008)] to approximate the free surface of a dam-break flow, when the free surface evolution is governed by the KdV equation. The KdV equation is obtained when dispersive effects are incorporated into the shallow water theory, a derivation of which is given in [Whitham (1974)] (p. 460). In [Leach and Needham (2008)], the method of matched asymptotic co-ordinate expansions was used to approximate the large time development of the solution to the dam-break problem for the KdV equation.

All the work referenced above is fundamentally based upon the shallow water approximation and dispersive modifications. In terms of the dam-break problem, the relevance of the nonlinear shallow water equations was investigated in [Dutykh and Mitsotakis (2010)], where the authors compared Navier-Stokes simulations with an analytical solution and numerical approximations of the nonlinear shallow water equations, with a dry bed downstream from the dam. The authors found that the Navier-Stokes simulations do not predict the parabolic behaviour in the free surface, as predicted by the analytical solution to the nonlinear shallow water equations. Their graphs, which depict the free surface behaviour for a small time, also suggests a difference in the wavefront structure between the two simulations. The nonlinear shallow water equations show the wave front being tangent to the bed, in agreement with [Stoker (1957)]. However, the Navier-Stokes simulations show something of a bore at the wavefront in the initial stage, which smooths out over time. The authors also noted that experimental data from the initial stages of dam-break, [Stansby *et al.* (1998)], also differs from the predictions of the shallow water equations. Thus, there is a case to address these discrepancies by perform-ing an analysis of the dam-break problem formulated from the full water wave equations. Henceforth we will refer to this as the full dam-break problem. Among the most notable works on the full dam-break problem is that of Pohle, [Pohle (2006)], and Stoker, [Stoker (1957)] (Chap. 12). In, [Stoker (1957)] (Chap. 12), the full dam-break problem was considered, and an approximate solution was constructed during the initial stages after the break. The equations of motion were assumed to have power series solutions in time, t, for the pressure and the particle displacement, and each were solved up to $O\left(t^2\right)$ as $t \to 0$, but it was noted by the author that these solutions become singular at the point where the dam meets the bed (at the wave front).

As far as we are aware, this is the first work that constructs the solution at the initial stage of the full dam-break problem. Hunt [Hunt (1983)] also constructed an asymptotic approximation to the full dam-break problem as $t \to 0$; however there appears to be little subsequent work on the full dam-break problem using this technique. This was also noted in [Korobkin and Yilmaz (2009)], where it is reported that "there are very few asymptotic analyses of the dam-break problem".

One paper which does take this approach, and is closely related to this monograph, is that by Korobkin and Yilmaz, [Korobkin and Yilmaz (2009)]. The authors consider the case where a vertical dam sits on a horizontal bed and separates a layer of fluid, which extends horizontally to infinity, from a region where there is no fluid. The authors construct a small time approximation to this dam-break problem by developing the solution to the fluid velocity potential and the free surface in a series expansion in the time variable. The problem is solved at each order and a second order outer solution is obtained. This outer solution is singular at the point where the dam meets the bed, (at the wave front) due to logarithmic singularities, which is the same point where the series solution in Stoker's work, [Stoker (1957)], became singular. In [Stoker (1957)], the author said of this singular point that "there would be turbulence and continuous breaking at the front of the wave anyway so that any solution ignoring these factors would be unrealistic for that part of the flow." The authors in [Korobkin and Yilmaz (2009)], however, do construct a leading order approximation in an inner region around this singular point. The authors show that the outer and inner region solutions asymptotically match according to Van Dyke's matching principle [Van Dyke (1975)], and that the inner region solution describes a jet formation that moves along the dry bed.

In this monograph, we consider in full detail a linearised two-dimensional full dam-break problem. We begin by formulating the linearised full dam-break problem, to which we apply the complex Fourier transform to obtain the exact solution for the fluid velocity potential and free surface in complex Fourier integral form. Uniform asymptotic approximations to the free surface are then obtained in detail for small time, in the far fields and for large time. As a supplement, the exact free surface solution is also computed numerically, with suitable error bounds, and the numerical solutions are then used to compare with the asymptotic forms in the small and large time limits. The numerical solutions also provide a (error controlled) link between the asymptotic form for small time and the asymptotic form for large time. In addition, we write down (trivially) the free surface solution to the corresponding linearised shallow water dam-break problem. We are then able to compare this directly with the asymptotic forms of the exact free surface solution to the linearised full dam-break problem in both the small and large time limits. We also make a comparison with the numerical solution of the linearised full dam-break problem.

In Chap. 2, we introduce and formulate the full dam-break problem. Here, we consider a two-dimensional dam-break problem, where the dam is inclined, and the fluid is initially at rest and at distinct constant depths either side of the dam. The fluid lies on a horizontal impermeable base and is bounded above by a free surface, as shown in Fig. (2.1). We assume that the dam disappears instantly at the initial time leading to a a subsequent dam-break flow. Our aim is to determine the solution for the free surface of the fluid and the fluid velocity field. We begin by formulating the dam-break problem for an incompressible and inviscid fluid. The fluid velocity field is governed by the conservation of mass (continuity equation) and conservation of momentum (Euler equations). The only force driving the flow is gravity, so that, as the fluid is initially at rest, the fluid velocity field is subsequently irrotational. This allows the introduction of a velocity potential, from which we obtain the governing equation for the velocity potential as Laplace's equation, with appropriate initial conditions and boundary conditions. We then non-dimensionalise this nonlinear problem with respect to the initial depth of the fluid layer forward of the dam. This nonlinear problem depends upon two parameters, the dimensionless step length and step height of the dam.

In Chap. 3, we consider the situation where the dam has a small dimensionless step height and an $O(1)$ step length. In this situation the full dam-break problem may be linearised to obtain the linearised full dam-break problem. We then apply complex Fourier transforms to the linearised full dam-break problem. Once the transformed problem is solved we apply the inverse Fourier transform to obtain exact solutions for the free surface and fluid velocity potential to the linearised full dam-break problem in integral forms. At this point we turn our attention to the detailed structure of the solution for the free surface.

In Chap. 4, we construct a uniform asymptotic approximation to the free surface evolution during the initial stages of the dam-break flow as time $t \to 0^+$. Expanding the integrand as $t \to 0^+$, the free surface solution emerges as an asymptotic series in t. It becomes clear from this expansion that the asymptotic structure is made up of five distinct asymptotic regions. There are two inner regions, which are small regions around the initial corners of the free surface, and there are three outer regions away from the initial corners. We first determine the asymptotic development of the free surface up to $O\left(t^2\right)$ in the outer regions, and then in the inner regions. We demonstrate that Van Dyke's asymptotic matching principle [Van Dyke (1975)] is satisfied accordingly. The results reveal an $O\left(t^2\right)$ evolution of the free surface in the outer regions, and an $O\left(t^2 \log t\right)$ evolution in each of the inner regions, where we uncover incipient jet formation and collapse. A qualitative sketch and graphs depicting the short time behaviour of the free surface are given, and the solution exhibits behaviour which accords with that reported in [Korobkin and Yilmaz (2009)], where jet formation is observed in a similar problem. This also agrees with the experimental data in [Stansby *et al.* (1998)], in which the authors describe incipient jet formation in the free surface at the wavefront.

In Chap. 5, we construct a uniform asymptotic approximation to the free surface in the far fields, as $|x| \to \infty$. This is achieved via a detailed application of the

method of steepest descents. It is established that the asymptotic approximation
in the far fields, as $|x| \to \infty$, consists of three distinct asymptotic regions, namely
$t = o(1)$, $t = O(1)$ and $t = [o(1)]^{-1}$ as $|x| \to \infty$. The asymptotic development
is derived in each region, and it is verified that asymptotic matching is satisfied
between each region according to Van Dyke's principle [Van Dyke (1975)]. It is
established that the free surface differs from the initial conditions by an exponen-
tially small order in x, as $|x| \to \infty$ in the far fields.

In Chap. 6, we construct a uniform asymptotic approximation to the free surface for large time, as $t \to \infty$. The detailed structure consists of five distinct asymptotic regions, three outer regions and two inner regions. We find an outer region
where the free surface oscillates, and this region connects to two inner regions
where the free surface is described by Airy functions and their integrals. Each
inner region then connects to a corresponding outer region which extends into the
far field. In these outer regions the free surface differs from the initial conditions
by an exponentially small order in t, as $t \to \infty$. It is also verified that asymptotic matching is satisfied between each region according to Van Dyke's principle
[Van Dyke (1975)]. A sketch detailing the asymptotic structure of the free surface
at a large time is given and graphs of the free surface are also presented.

In Chap. 7, we give a detailed summary of the exact solution and the asymptotic structure obtained in Chaps. 3–6. Sketches and graphs are given to illustrate
the detailed free surface structure for $t \to 0$ and $t \to \infty$.

In Chap. 8, we perform a numerical evaluation of the exact free surface solution obtained in Chap. 3. We employ Simpson's rule and give a precise error
bound. The numerical evaluation is performed for various times and graphs for
each time are shown. Comparisons are made with the asymptotic approximations,
and excellent agreement is observed.

In Chap. 9, we consider the problem where the free surface, with the same
initial conditions as in Chap. 3, is governed by the linear wave equation associated
with the linearised shallow water theory. The problem is solved by D'Alembert's
general solution, and graphs are given for various times. Sketches of the solution
are given and illustrate the free surface behaviour for $t \to 0^+$ and $t \to \infty$. A detailed
comparison is then made between the solution to the linearised shallow water
theory, and the solution to the linearised full theory. This involves comparisons,
in particular, as $t \to 0^+$ and as $t \to \infty$.

Finally it is instructive to compare the results in this monograph to experimental studies performed on dam-break problems. The dam-break problem has been
analysed experimentally a number of times. Most relevant to this monograph is
the experimental study by Stansby, Chegini and Barnes, [Stansby *et al.* (1998)].
Here, the authors set up a flume with a horizontal base, and had a metal plate

acting as a dam. Here water was initially at rest, at different depths, either side of the dam, and the dam was then rapidly drawn vertically upwards out of the water via a pulley. The subsequent flow was visualised by a laser light sheet and recorded on video camera. In the short time they observed a jet formation at the base of the dam. This is very similar behaviour to that which is predicted in this monograph, where after a short time the asymptotic and numerical approximations show an incipient jet formation and collapse in the free surface in regions around the initial corners of the free surface. The authors also compared their experimental results to analytical results of the nonlinear shallow water equations. Here, the authors found that the experimental and analytical results had a close agreement after a certain period of time, however there was more significant difference between the two during the initial stages after the dam-break. We find similar behaviour in this monograph where the full linearised theory and the linearised shallow water theory, as both $t \to 0^+$ and $t \to \infty$, agree at leading order in all outer asymptotic regions, but may differ at $O(1)$ in thin inner regions located at the upstream and downstream transition waterfronts.

Finally, we observe that the methodology developed in this monograph to analyse the complex Fourier integrals arising in the exact solution to the linearised dam-break problem, is, in many parts, novel, and will have applications in other linear wave evolution problems. Specifically, we anticipate that the detailed theory developed in the monograph for a linear ramp dam profile will be readily generalised for alternative dam profiles, and the main conclusions will remain unchanged qualitatively at least.

Chapter 2

Formulation of the Dam-Break Problem

Throughout this monograph we consider a fluid which is incompressible and inviscid. Further, the only external force acting on the fluid is gravity. Specifically, we consider the situation when a body of fluid is initially at rest above an impermeable rigid and horizontal boundary, and is bounded above by a free surface, which initially is stationary, and represents a transition from one uniform depth to another uniform depth. The initial displacement of the free surface varies only in one horizontal space dimension. This problem is generally referred to as the two-dimensional dam-break problem. The spatial domain is specified by the Cartesian coordinate system (x, z), with z meaning distance vertically upwards and x meaning distance horizontally, whilst $t \geq 0$ represents time. A fixed position in the spatial domain will be denoted as $\underline{r} = (x, z) = x\underline{i} + z\underline{k}$, where $\underline{i} = (1, 0)$ and $\underline{k} = (0, 1)$ represent unit vectors in the x and z directions respectively. The velocity field of the fluid is denoted as $\underline{u} = (u(x, z, t), w(x, z, t)) = u(x, z, t)\underline{i} + w(x, z, t)\underline{k}$, the free surface of the fluid is located at $z = \eta(x, t)$ and the impermeable base is set at $z = -h_0$. The region occupied by the fluid at time $t \geq 0$ will be denoted as $D(t)$, where

$$D(t) = \{(x, z) : (x, z) \in \mathbb{R} \times (-h_0, \eta(x, t))\},$$

with the closure of the region denoted $\bar{D}(t)$, where

$$\bar{D}(t) = \{(x, z) : (x, z) \in \mathbb{R} \times [-h_0, \eta(x, t)]\}.$$

The initial region occupied by the fluid, as illustrated in Fig. (2.1), is denoted as $D(0) = D_0$, with closure $\bar{D}(0) = \bar{D}_0$. The initial displacement of the free surface is given as

$$z = \eta(x, 0) = \eta_0(x) = \begin{cases} 0 & \text{when } x \geq l_0, \\ -\dfrac{\alpha h_0}{l_0} x + \alpha h_0 & \text{when } 0 < x < l_0, \\ \alpha h_0 & \text{when } x \leq 0, \end{cases} \tag{2.1}$$

with $l_0 > 0$ and $\alpha \geq 0$.

9

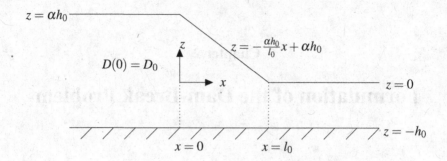

Fig. 2.1: Initial displacement of the fluid layer.

As the fluid is initially at rest, it follows from Kelvin's Circulation Theorem that the fluid velocity field is irrotational for all $t \geq 0$. We can now introduce a fluid velocity potential $\phi = \phi(x,z,t)$, so that $\underline{u} = \nabla \phi$. The initial velocity field at $t = 0$ is $\underline{u} = \nabla \phi = \underline{0}$. Therefore the initial velocity potential may be taken as

$$\phi(x,z,0) = 0, \quad (x,z) \in \bar{D}_0. \tag{2.2}$$

The fluid is released from rest in its initial configuration and the subsequent fluid velocity field, $\underline{u}$, must satisfy the continuity equation,

$$\nabla \cdot \underline{u} = 0, \quad (x,z) \in D(t), \quad t > 0,$$

which, in terms of the fluid velocity potential becomes

$$\nabla^2 \phi = 0, \quad (x,z) \in D(t), \quad t > 0, \tag{2.3}$$

which is the governing equation for the fluid velocity potential. The fluid pressure field, $p = p(x,z,t)$ (relative to atmospheric pressure p_A), is then given by the unsteady Bernoulli equation,

$$\phi_t + \frac{1}{2}|\nabla \phi|^2 + \frac{p}{\rho} + gz = 0, \quad \text{with } (x,z) \in D(t), \quad t > 0. \tag{2.4}$$

We now consider boundary conditions on the free surface $z = \eta(x,t)$. The fluid pressure on the free surface is $p(x, \eta(x,t),t) = 0$, which, via the Bernoulli equation (2.4) gives the boundary condition,

$$\phi_t + \frac{1}{2}|\nabla \phi|^2 + g\eta = 0, \quad \text{on } z = \eta(x,t), \ (x,t) \in \mathbb{R} \times \mathbb{R}^+, \tag{2.5}$$

which is the dynamic free surface boundary condition. In addition, we require that the normal fluid velocity at the free surface is equal to the normal displacement velocity of the free surface. This gives,

$$\eta_t = \phi_z - \eta_x \phi_x, \quad \text{on } z = \eta(x,t), \ (x,t) \in \mathbb{R} \times \mathbb{R}^+, \tag{2.6}$$

which is the kinematic free surface boundary condition. There is also the boundary condition on the impermeable boundary, where there is zero normal fluid velocity at the boundary, that is

$$\nabla \phi \cdot \underline{k} = 0, \quad \text{at } z = -h_0, \ (x,t) \in \mathbb{R} \times \mathbb{R}^+. \tag{2.7}$$

The far field boundary conditions on the free surface are

$$\eta(x,t) \to \begin{cases} 0, & \text{as } x \to \infty \quad \text{with } t \geq 0, \\ \alpha h_0, & \text{as } x \to -\infty \quad \text{with } t \geq 0. \end{cases} \tag{2.8}$$

The fluid starts from rest, therefore the far field conditions for the velocity potential are $|\nabla \phi| \to 0$, as $|x| \to \infty$ uniformly in $\bar{D}(t), t \geq 0$. On using (2.5) this may be written as

$$\phi_t(x,z,t) \to \begin{cases} 0, & \text{as } x \to \infty \quad \text{with } (z,t) \in [-h_0, 0] \times [0, \infty), \\ -g\alpha h_0, & \text{as } x \to -\infty \quad \text{with } (z,t) \in [-h_0, \alpha h_0] \times [0, \infty), \end{cases}$$

which leads to

$$\phi(x,z,t) \to \begin{cases} C_+, & \text{as } x \to \infty \quad \text{with } (z,t) \in [-h_0, 0] \times [0, \infty), \\ -g\alpha h_0 t + C_-, & \text{as } x \to -\infty \quad \text{with } (z,t) \in [-h_0, \alpha h_0] \times [0, \infty), \end{cases}$$

where $C_+, C_- \in \mathbb{R}$ are constants. Initial condition (2.2) then requires $C_+ = C_- = 0$ to give the far field conditions for ϕ as

$$\phi(x,z,t) \to \begin{cases} 0, & \text{as } x \to \infty \quad (z,t) \in [-h_0, 0] \times [0, \infty), \\ -g\alpha h_0 t, & \text{as } x \to -\infty \quad (z,t) \in [-h_0, \alpha h_0] \times [0, \infty). \end{cases} \tag{2.9}$$

The dam-break problem is governed by (2.3) with the conditions (2.1), (2.2), (2.5), (2.6), (2.7), (2.8) and (2.9). Regularity conditions on ϕ and η require the following;

$$\phi(x,z,t) \in C(\bar{D}_\infty) \cap C^{1,1,1}(D_\infty \cup \partial D_\infty) \cap C^{2,2,0}(D_\infty),$$

$$\eta(x,t) \in C(\mathbb{R} \times [0,\infty)) \cap C^{1,1}(\mathbb{R} \times (0,\infty)),$$

with

$$D_\infty = \{(x,z,t) : (x,z) \in D(t), \ t \in (0,\infty)\},$$

$$\partial D_\infty = \{(x,z,t) : (x,z) \in \bar{D}(t) \backslash D(t), \ t \in (0,\infty)\}.$$

The problem is non-dimensionalised with respect to the depth scale, h_0, so that the step length l_0 can be written as $l_0 = \beta h_0$ with the parameter $\beta(>0) \in \mathbb{R}$. The following scales are used to non-dimensionalise the problem, namely,

$$x = x'h_0, \ z = z'h_0, \ \eta = \eta'h_0, \ \phi = \phi'h_0\sqrt{gh_0}, \ t = t'\sqrt{\tfrac{h_0}{g}}, \tag{2.10}$$

where the dashes represent the dimensionless variables. Substituting (2.10) into the dimensional problem we obtain the dimensionless dam-break problem given below:

$$\nabla^2 \phi = 0 \quad \text{for } (x,z) \in D(t), t > 0. \tag{2.11}$$

$$\nabla \phi \cdot \underline{k} = 0 \quad \text{at } z = -1, \text{ with } (x,t) \in \mathbb{R} \times \mathbb{R}^+. \tag{2.12}$$

$$\eta_t + \eta_x \phi_x - \phi_z = 0 \quad \text{at } z = \eta(x,t), (x,t) \in \mathbb{R} \times \mathbb{R}^+. \tag{2.13}$$

$$\phi_t + \frac{1}{2}\left(\phi_x^2 + \phi_z^2\right) + \eta = 0 \quad \text{at } z = \eta(x,t), (x,t) \in \mathbb{R} \times \mathbb{R}^+. \tag{2.14}$$

$$\phi(x,z,t) \to 0 \quad \text{as } x \to \infty \text{ uniformly in } \bar{D}(t) \text{ with } t \geq 0. \tag{2.15}$$

$$\phi(x,z,t) \to -\alpha t \quad \text{as } x \to -\infty \text{ uniformly in } \bar{D}(t) \text{ with } t \geq 0. \tag{2.16}$$

$$\eta(x,t) \to \begin{cases} 0 & \text{as } x \to \infty \text{ with } t \geq 0, \\ \alpha & \text{as } x \to -\infty \text{ with } t \geq 0. \end{cases} \tag{2.17}$$

$$\phi(x,z,0) = 0 \quad \text{for } (x,z) \in \bar{D}(0). \tag{2.18}$$

$$\eta(x,0) = \eta_0(x) \quad \text{for } x \in \mathbb{R}, \tag{2.19}$$

where

$$\eta_0(x) = \begin{cases} 0 & \text{when } x \geq \beta, \\ \alpha\left(1 - \frac{1}{\beta}x\right) & \text{when } 0 < x < \beta, \\ \alpha & \text{when } x \leq 0, \end{cases}$$

together with the regularity conditions,

$$\phi(x,z,t) \in C(\bar{D}_\infty) \cap C^{1,1,1}(D_\infty \cup \partial D_\infty) \cap C^{2,2,0}(D_\infty),$$

$$\eta(x,t) \in C(\mathbb{R} \times [0,\infty)) \cap C^{1,1}(\mathbb{R} \times (0,\infty)).$$

We now address problems (2.11)–(2.19). We will refer to problems (2.11)–(2.19) as [IBVP]. We begin by considering a linearised form of [IBVP].

Chapter 3

The Linearised Dam-Break Problem

In this chapter we consider a linearised form of [IBVP], which corresponds to a dam with a small step height and slope. To this end, we introduce appropriate scalings for $\phi(x,z,t)$ and $\eta(x,t)$ from which we formulate the linearised dam-break problem. Applying the Fourier transform to our linearised problem, we obtain exact solutions for the scaled fluid velocity potential $\bar{\phi}(x,z,t)$ and scaled free surface displacement $\bar{\eta}(x,t)$.

3.1 Formulation of the Linearised Problem

Consider [IBVP] for a small step height and slope, that is for $0 \le \alpha \ll 1$ and $\beta = O(1)$ as $\alpha \to 0$. In the case that $\alpha = 0$ the solution to [IBVP] is

$$\phi(x,z,t) = 0, \quad \text{for } (x,z) \in \bar{D}(t), \quad t \ge 0.$$

$$\eta(x,t) = 0, \quad x \in \mathbb{R}, \, t \ge 0.$$

For $0 < \alpha \ll 1$, we write

$$\phi = \alpha\bar{\phi}, \quad \eta = \alpha\bar{\eta}, \tag{3.1}$$

with $\bar{\phi}, \bar{\eta} = O(1)$ as $\alpha \to 0$. On substituting (3.1) into [IBVP], and neglecting terms of $O(\alpha^2)$, we obtain the following linearised dam-break problem, namely

$$\nabla^2\bar{\phi} = 0 \quad \text{for } (x,z,t) \in \mathbb{R} \times (-1,0) \times \mathbb{R}^+. \tag{3.2}$$

$$\bar{\phi}_z = 0 \quad \text{at } z = -1, \text{ with } (x,t) \in \mathbb{R} \times \mathbb{R}^+. \tag{3.3}$$

$$\bar{\eta}_t - \bar{\phi}_z = 0 \quad \text{at } z = 0, \text{ with } (x,t) \in \mathbb{R} \times \mathbb{R}^+. \tag{3.4}$$

$$\bar{\phi}_t + \bar{\eta} = 0 \quad \text{at } z = 0, \text{ with } (x,t) \in \mathbb{R} \times \mathbb{R}^+. \tag{3.5}$$

13

$$\bar{\phi}(x,z,t) \to 0 \quad \text{as } x \to \infty \text{ uniformly for } z \in [-1,0], \quad t \geq 0. \tag{3.6}$$

$$\bar{\phi}(x,z,t) \to -t \quad \text{as } x \to -\infty \text{ uniformly for } z \in [-1,0], \quad t \geq 0. \tag{3.7}$$

$$\bar{\eta}(x,t) \to \begin{cases} 0 & \text{as } x \to \infty \, t \geq 0, \\ 1 & \text{as } x \to -\infty \, t \geq 0, \end{cases} \tag{3.8}$$

with the initial conditions

$$\bar{\phi}(x,z,0) = 0 \quad \text{for } (x,z) \in \mathbb{R} \times [-1,0], \tag{3.9}$$

$$\bar{\eta}(x,0) = \bar{\eta}_0(x) \quad \text{for } x \in \mathbb{R}, \tag{3.10}$$

where

$$\bar{\eta}_0(x) = \begin{cases} 0, & x \geq \beta, \\ \dfrac{1}{\beta}(\beta - x), & 0 < x < \beta, \\ 1, & x \leq 0. \end{cases}$$

The following regularity conditions on $\bar{\phi}$ and $\bar{\eta}$ are also required:

$$\bar{\phi} \in C(\mathbb{R} \times [-1,0] \times \bar{\mathbb{R}}^+) \cap C^{1,1,1}(\mathbb{R} \times [-1,0] \times \mathbb{R}^+) \cap C^{2,2,0}(\mathbb{R} \times (-1,0) \times \mathbb{R}^+), \tag{3.11}$$

$$\bar{\eta} \in C(\mathbb{R} \times \bar{\mathbb{R}}^+) \cap C^{1,1}(\mathbb{R} \times \mathbb{R}^+). \tag{3.12}$$

We will refer to the linearised problems (3.2)–(3.12) as [LIBVP].

It is our intention to address the solution to [LIBVP] via the theory of complex Fourier transforms. To this end, we anticipate that the far field boundary conditions (3.6), (3.7) and (3.8) are achieved through terms exponentially small in x as $|x| \to \infty$. That is, we have

$$\bar{\eta}(x,t) \sim \begin{cases} O(\exp(-\lambda_+ x)) & \text{as } x \to \infty, t \geq 0, \\ 1 + O(\exp(\lambda_- x)) & \text{as } x \to -\infty, t \geq 0. \end{cases} \tag{3.13}$$

$$\bar{\phi}(x,z,t) \sim \begin{cases} O(\exp(-\lambda_+ x)) & \text{as } x \to \infty \text{ uniformly for } z \in [-1,0], t \geq 0, \\ -t + O(\exp(\lambda_- x)) & \text{as } x \to -\infty \text{ uniformly for } z \in [-1,0], t \geq 0. \end{cases} \tag{3.14}$$

where λ_+, λ_- are constants.

3.2 Exact Solution to the Linearised Problem

Now let $\bar{\phi} : \bar{D}_\infty \to \mathbb{R}$ and $\bar{\eta} : \mathbb{R} \times [0,\infty) \to \mathbb{R}$ be a solution to [LIBVP]. Then the regularity conditions (3.11), (3.12) and the decay estimates (3.13), (3.14) allow us to introduce the Fourier transforms

$$\hat{\eta}(k,t) = \int_{-\infty}^{\infty} \bar{\eta}(x,t) \exp(ikx)\, dx, \quad (k,t) \in D \times \bar{\mathbb{R}}^+,$$

$$\hat{\phi}(k,z,t) = \int_{-\infty}^{\infty} \bar{\phi}(x,z,t) \exp(ikx)\, dx, \quad (k,z,t) \in D \times [-1,0] \times \bar{\mathbb{R}}^+,$$

where $D \subset \mathbb{C}$ is the strip $D = \{k = \sigma + i\tau : \sigma \in (-\infty,\infty), \tau \in (-\lambda_+, 0)\}$. Via the regularity conditions (3.11) and (3.12), together with the far field boundary conditions (3.13) and (3.14), it follows that $\hat{\eta} : D \times \bar{\mathbb{R}}^+ \to \mathbb{C}$ is such that $\hat{\eta} \in C(D \times \bar{\mathbb{R}}^+)$ and is analytic in k for all $(k,t) \in D \times \bar{\mathbb{R}}^+$. Similarly, $\hat{\phi} : D \times [-1,0] \times \bar{\mathbb{R}}^+ \to \mathbb{C}$ is such that $\hat{\phi} \in C(D \times [-1,0] \times \bar{\mathbb{R}}^+)$ and is analytic in k for all $(k,z,t) \in D \times [-1,0] \times \bar{\mathbb{R}}^+$. Applying the Fourier transform to [LIBVP] we arrive at the following problem for the transformed variables $\hat{\eta}$ and $\hat{\phi}$, namely,

$$\hat{\phi}_{zz} - k^2\hat{\phi} = 0, \quad (k,z,t) \in D \times (-1,0) \times \mathbb{R}^+. \tag{3.15}$$

$$\hat{\phi}_z = 0 \quad \text{at } z = -1 \text{ for } (k,t) \in D \times \mathbb{R}^+. \tag{3.16}$$

$$\hat{\eta}_t - \hat{\phi}_z = 0 \quad \text{at } z = 0 \text{ for } (k,t) \in D \times \mathbb{R}^+. \tag{3.17}$$

$$\hat{\phi}_t + \hat{\eta} = 0 \quad \text{at } z = 0 \text{ for } (k,t) \in D \times \mathbb{R}^+. \tag{3.18}$$

$$\hat{\phi}(k,z,0) = 0, \quad (k,z) \in D \times [-1,0]. \tag{3.19}$$

$$\hat{\eta}(k,0) = \frac{1}{\beta k^2}(1 - \exp(ik\beta), \quad k \in D, \tag{3.20}$$

whilst the far field conditions, (3.13) and (3.14), require $\hat{\eta}$ and $\hat{\phi}$ to be analytic in k, for $k \in D$. From (3.15) we have

$$\hat{\phi}(k,z,t) = A(k,t)\cosh k(z+1) + B(k,t)\sinh k(z+1), \quad (k,z,t) \in D \times [-1,0] \times \bar{\mathbb{R}}^+,$$

with $A(k,t)$, $B(k,t)$ arbitrary functions of $(k,t) \in D \times \bar{\mathbb{R}}^+$. Applying (3.16) we obtain $B(k,t) = 0$ and hence

$$\hat{\phi}(k,z,t) = A(k,t)\cosh k(z+1), \quad (k,z,t) \in D \times [-1,0] \times \bar{\mathbb{R}}^+. \tag{3.21}$$

Now (3.17) and (3.18) require

$$\hat{\eta}_t - kA(k,t)\sinh k = 0,$$

$$A_t \cosh k + \hat{\eta} = 0,$$

for $(k,t) \in D \times \mathbb{R}^+$, subject to the initial conditions from (3.19) and (3.20),

$$A(k,0) = 0,$$

$$\hat{\eta}(k,0) = \frac{1}{\beta k^2}(1 - \exp(ik\beta)),$$

for $k \in D$. These give us

$$\left. \begin{array}{l} A(k,t) = -\dfrac{(1 - \exp(ik\beta))\sin(\gamma(k)t)}{\beta k^2 \gamma(k) \cosh k} \\[3mm] \hat{\eta}(k,t) = \dfrac{(1 - \exp(ik\beta))}{\beta k^2}\cos(\gamma(k)t) \end{array} \right\} (k,t) \in D \times \mathbb{R}^+, \qquad (3.22)$$

with $\gamma^2(k) = k \tanh k$. Finally we obtain, via (3.21) and (3.22),

$$\hat{\phi}(k,z,t) = -\frac{(1 - \exp(ik\beta))\sin(\gamma(k)t)}{\beta k^2 \gamma(k)\cosh k}\cosh k(z+1), \quad (k,z,t) \in D \times [-1,0] \times \mathbb{R}^+.$$

To preserve analyticity of $\hat{\eta}$ and $\hat{\phi}$ for $k \in D$ it is required that $\gamma(k)$ is also analytic in D. Thus we must have $\lambda_+ = \frac{\pi}{2}$ and set $\gamma(k) = (k \tanh k)^{\frac{1}{2}}$ with branch-cuts in the k-plane, as shown in Fig. (3.1), and $\arg(\gamma(k)) = 0$ for $k \in \mathbb{R}^+$. We observe that $\gamma(k)$ is analytic in the cut k-plane, and,

$$\gamma^2(-k) = \gamma^2(k)$$

throughout the k-plane, whilst

$$\gamma(-k) = -\gamma(k) \qquad (3.23)$$

throughout the cut k-plane. Also,

$$\gamma(k) = k + O(k^3) \quad \text{as } k \to 0 \qquad (3.24)$$

and

$$\gamma(k) \sim \begin{cases} k^{\frac{1}{2}} & \text{as } |k| \to \infty \text{ on } Re(k) > 0, \\[2mm] -(-k)^{\frac{1}{2}} & \text{as } |k| \to \infty \text{ on } Re(k) < 0, \end{cases} \qquad (3.25)$$

with the principal branch of the square root implied. In addition $\gamma^2(k)$ has simple poles at $k = \pm k_n$ with

$$k_n = i\left(n - \frac{1}{2}\right)\pi, \quad n = 1,2,\dots$$

and

$$Res(\gamma^2(k), k_n) = i\left(n - \frac{1}{2}\right)\pi, \quad n = 1,2,\dots,$$

$$Res(\gamma^2(k), -k_n) = -i\left(n - \frac{1}{2}\right)\pi, \quad n = 1, 2, \ldots.$$

We observe that $\hat{\eta}(k,t)$ and $\hat{\phi}(k,z,t)$ are analytic for $k \in D$ as required, each having a simple pole at $k = 0$ in the cut k-plane. Moreover

$$\hat{\eta}(k,t) = -\frac{i}{k} + O(1) \quad \text{as } k \to 0, \tag{3.26}$$

$$\hat{\phi}(k,z,t) = -\frac{it}{k} + O(1) \quad \text{as } k \to 0. \tag{3.27}$$

Also, for $k \in D$,

$$\hat{\eta}(k,t) = O\left(\frac{1}{|k|^2}\right), \quad \text{as } |k| \to \infty, \tag{3.28}$$

$$\hat{\phi}(k,t) = O\left(\frac{\exp(|k|z)}{|k|^{\frac{5}{2}}}\right), \quad \text{as } |k| \to \infty. \tag{3.29}$$

We can now apply the Fourier Inversion Theorem to recover $\bar{\eta}(x,t)$ and $\bar{\phi}(x,z,t)$ as

$$\bar{\eta}(x,t) = \frac{1}{2\pi\beta}\int_C \frac{1}{k^2}(1 - \exp(i\beta k))\cos(\gamma(k)t)\exp(-ixk)\,dk,$$

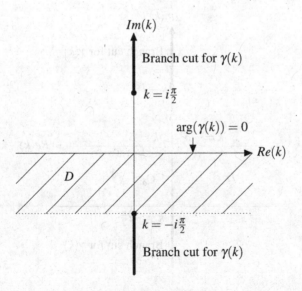

Fig. 3.1: Cut k-plane.

$$\bar{\phi}(x,z,t) = -\frac{1}{2\pi\beta} \int_C (1 - \exp(i\beta k)) \sin(\gamma(k)t) \frac{\cosh k(z+1)}{\gamma(k)k^2 \cosh k} \exp(-ixk) \, dk,$$

where C is any continuous and piecewise smooth contour in D from $Re(k) = -\infty$ to $Re(k) = +\infty$. An application of the Cauchy Residue Theorem with the estimates (3.28) and (3.29) allows us to choose C as the real k-axis, indented below the origin by a semi-circle of radius $0 < \delta \ll 1$, which we denote as C_δ, as shown in Fig. (3.2). Thus

$$\bar{\eta}(x,t) = \frac{1}{2\pi\beta} \int_{C_\delta} \frac{1}{k^2} (1 - \exp(i\beta k)) \cos(\gamma(k)t) \exp(-ixk) \, dk, \quad (x,t) \in \mathbb{R} \times \bar{\mathbb{R}}^+,$$
(3.30)

$$\bar{\phi}(x,z,t) = -\frac{1}{2\pi\beta} \int_{C_\delta} (1 - \exp(i\beta k)) \sin(\gamma(k)t) \frac{\cosh k(z+1)}{\gamma(k)k^2 \cosh k} \exp(-ixk) \, dk,$$
(3.31)

with $(x,z,t) \in \mathbb{R} \times [-1,0] \times \bar{\mathbb{R}}^+$.

We note that standard uniform convergence results, together with estimates (3.26) and (3.27) and integration by parts, establish that $\bar{\eta}$ and $\bar{\phi}$ given in (3.30) and (3.31) satisfy the regularity requirements (3.11) and (3.12).

In what follows it is convenient to write

$$\bar{\eta}(x,t) = \frac{1}{2\pi\beta} (I(x,t) - I(x-\beta,t)), \quad (x,t) \in \mathbb{R} \times \bar{\mathbb{R}}^+ \qquad (3.32)$$

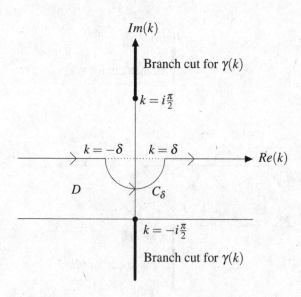

Fig. 3.2: Contour C_δ in the k-plane.

with

$$I(x,t) = \int_{C_\delta} \frac{1}{k^2} \cos(\gamma(k)t) \exp(-ikx)\, dk, \quad (x,t) \in \mathbb{R} \times \bar{\mathbb{R}}^+ \tag{3.33}$$

which has regularity $I \in C(\mathbb{R} \times \bar{\mathbb{R}}^+) \cap C^{1,1}(\mathbb{R} \times \mathbb{R}^+)$. Moreover, we have

$$I(-x,t) = -\int_{C_\delta^-} \frac{1}{k^2} \cos(\gamma(k)t) \exp(-ikx)\, dk, \quad (x,t) \in \mathbb{R} \times \bar{\mathbb{R}}^+$$

with C_δ^- shown in Fig. (3.3). Thus

$$I(x,t) - I(-x,t) = \int_{C_\delta} \frac{1}{k^2} \cos(\gamma(k)t) \exp(-ikx)\, dk + \int_{C_\delta^-} \frac{1}{k^2} \cos(\gamma(k)t) \exp(-ikx)\, dk$$

with $(x,t) \in \mathbb{R} \times \bar{\mathbb{R}}^+$. An application of the Cauchy Residue Theorem then gives

$$I(-x,t) = I(x,t) - 2\pi x \quad \forall (x,t) \in \mathbb{R} \times \bar{\mathbb{R}}^+. \tag{3.34}$$

Similarly, we write

$$\bar{\phi}(x,z,t) = \frac{1}{2\pi\beta} \left(J(x-\beta,z,t) - J(x,z,t) \right), \quad (x,z,t) \in \mathbb{R} \times [-1,0] \times \bar{\mathbb{R}}^+$$

with

$$J(x,z,t) = \int_{C_\delta} \frac{\cosh k(z+1) \sin(\gamma(k)t)}{\gamma(k)k^2 \cosh k} \exp(-ikx)\, dk, \quad (x,z,t) \in \mathbb{R} \times [-1,0] \times \bar{\mathbb{R}}^+$$

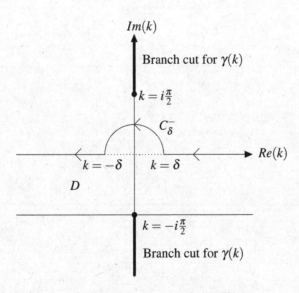

Fig. 3.3: Contour C_δ^- in the k-plane.

and which has regularity $J(x,z,t) \in C(\mathbb{R} \times [-1,0] \times \bar{\mathbb{R}}^+) \cap C^{1,1,1}(\mathbb{R} \times [-1,0] \times \mathbb{R}^+) \cap C^{2,2,0}(\mathbb{R} \times (-1,0) \times \mathbb{R}^+)$. It is also readily established that

$$J(-x,z,t) = J(x,z,t) - 2\pi xt, \quad \forall (x,z,t) \in \mathbb{R} \times [-1,0] \times \bar{\mathbb{R}}^+.$$

We now investigate the coordinate expansions for $\bar{\eta}(x,t)$ as given by (3.30) in the cases

(a) $t \to 0$ uniformly for $x \in [-X, X]$, $X \geq 0$.
(b) $|x| \to \infty$ uniformly for $t \in [0, T]$, $T \geq 0$.
(c) $t \to \infty$ uniformly for $x \in \mathbb{R}$.

Chapter 4

Coordinate Expansions for $\bar{\eta}(x,t)$ as $t \to 0$

In this chapter we consider the free surface displacement $\bar{\eta}(x,t)$, as given by (3.30), for $x \in \mathbb{R}$ as $t \to 0$. We begin by approximating $I(x,t)$ in (3.33) as $t \to 0$ when $x \geq 0$, and then construct the approximation to $\bar{\eta}(x,t)$, for $x \in \mathbb{R}$ as $t \to 0$, via (3.32) and (3.34). We obtain an approximation to $\bar{\eta}(x,t)$ for $x \in \mathbb{R}$ as $t \to 0$ which consists of an outer region and two inner regions, which are $O\left(t^2\right)$ neighbourhoods of the corners in the initial data, at $x = 0$ and $x = \beta$. The inner regions show incipient jet formation near $x = \beta$ and incipient collapse near $x = 0$.

4.1   Outer Region Coordinate Expansion for $I(x,t)$ as $t \to 0$

Consider $I(x,t)$, as given in (3.33), as $t \to 0$, when $x \geq 0$. First write,

$$I(x,t) = \int_{C_\delta^{\theta(t)}} \frac{1}{k^2} \cos(\gamma(k)t) \exp(-ikx)\,dk + \int_{\theta(t)}^{\infty} \frac{1}{k^2} \cos(\gamma(k)t) \exp(-ikx)\,dk$$
$$+ \int_{-\infty}^{-\theta(t)} \frac{1}{k^2} \cos(\gamma(k)t) \exp(-ikx)\,dk, \tag{4.1}$$

where $\theta(t) \to \infty$ as $t \to 0$, and $\theta(t) = o\left(t^{-2}\right)$ as $t \to 0$. Here $C_\delta^{\theta(t)}$ is that part of the contour C_δ between $k = -\theta(t)$ and $k = \theta(t)$, as shown in Fig. (4.1). It is straightforward to estimate, *when $x > 0$*, that,

$$\int_{C_\delta^{\theta(t)}} \frac{1}{k^2} \cos(\gamma(k)t) \exp(-ikx)\,dk = a_0(x) + t^2 a_1(x) + O\left(\frac{1}{x\theta(t)^2}, \frac{t^2}{x\theta(t)}, \theta(t)t^4\right) \tag{4.2}$$

as $t \to 0$, where $a_0(x)$ and $a_1(x)$, for $x \in \mathbb{R}^+$, are given by (4.6) and (4.7). In addition, we have, after an integration by parts, *when $x > 0$*, that,

$$\left| \int_{\theta(t)}^{\infty} \frac{1}{k^2} \cos(\gamma(k)t) \exp(-ikx)\,dk \right|, \left| \int_{-\infty}^{-\theta(t)} \frac{1}{k^2} \cos(\gamma(k)t) \exp(-ikx)\,dk \right| \leq \frac{2}{x\theta(t)^2} \tag{4.3}$$

21

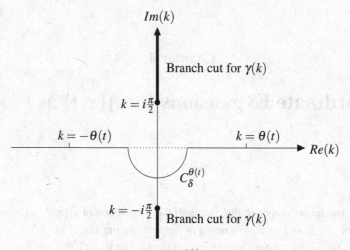

Fig. 4.1: Contour $C_\delta^{\theta(t)}$ in the k-plane.

as $t \to 0$. For any $\varepsilon > 0$, we may take $\theta(t) = t^{-2+\varepsilon}$, after which (4.1)–(4.3) gives,

$$I(x,t) = a_0(x) + t^2 a_1(x) + O\left(\frac{t^{4-2\varepsilon}}{x}, \frac{t^{4-\varepsilon}}{x}, t^{2+\varepsilon}\right) \qquad (4.4)$$

as $t \to 0$ with $x > 0$. Clearly, the approximation in (4.4) fails when $x = O(t^2)$ as $t \to 0$. Thus, we have from (4.4),

$$I(x,t) = a_0(x) + t^2 a_1(x) + o(t^2) \qquad (4.5)$$

as $t \to 0$ uniformly for $x \in \bar{\mathbb{R}}^+ \setminus [0, \Delta(t))$, where $\Delta(t) = O(t^2)$ as $t \to 0$. Here

$$a_0(x) = \int_{C_\delta} \frac{1}{k^2} \exp(-ikx)\,dk, \quad x \in \mathbb{R}^+, \qquad (4.6)$$

$$a_1(x) = -\frac{1}{2} \int_{C_\delta} \frac{\tanh k}{k} \exp(-ikx)\,dk, \quad x \in \mathbb{R}^+. \qquad (4.7)$$

We observe that $a_0, a_1 \in C^1(\mathbb{R}^+)$ via integration by parts and uniform convergence results. We now obtain explicit expressions for $a_0(x)$ and $a_1(x)$ for $x \in \mathbb{R}^+$. We begin with $a_0(x)$ for $x \in \mathbb{R}^+$.

Consider the integral in (4.6). To evaluate this integral, we integrate around the contour shown in Fig. (4.2). The integrand in (4.6) is analytic in the region enclosed by $C_\delta^R \cup C_R$. Hence via the Cauchy Residue Theorem,

$$\int_{C_\delta^R} \frac{\exp(-ikx)}{k^2}\,dk + \int_{C_R} \frac{\exp(-ikx)}{k^2}\,dk = 0. \qquad (4.8)$$

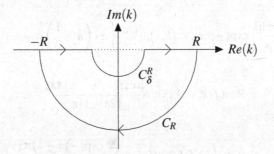

Fig. 4.2: Contour $C_\delta^R \cup C_R$ in the k plane.

On C_R, k can be written as

$$k = R\cos\theta + iR\sin\theta \quad \text{for } \theta \in [-\pi, 0],$$

so that

$$\int_{C_R} \frac{\exp(-ikx)}{k^2}\, dk = -\int_{\theta=-\pi}^0 \frac{R(-\sin\theta + i\cos\theta)}{R^2(\cos\theta + i\sin\theta)^2} \exp(-ixR(\cos\theta + i\sin\theta))\, d\theta.$$

Taking the absolute value gives

$$\left| \int_{C_R} \frac{\exp(-ikx)}{k^2}\, dk \right| \le \int_{\theta=-\pi}^0 \frac{1}{R} \exp(xR\sin\theta)\, d\theta.$$

Since $x > 0$, then $0 < \exp(xR\sin\theta) \le 1$, and so

$$\left| \int_{C_R} \frac{\exp(-ikx)}{k^2}\, dk \right| \le \int_{\theta=-\pi}^0 \frac{1}{R}\, d\theta$$

$$= \frac{\pi}{R} \to 0 \quad \text{as } R \to \infty.$$

Therefore, letting $R \to \infty$ in (4.8) we obtain

$$a_0(x) = \int_{C_\delta} \frac{\exp(-ikx)}{k^2}\, dk = 0, \quad x \in \mathbb{R}^+. \tag{4.9}$$

We now consider $a_1(x)$ for $x \in \mathbb{R}^+$. Since $x > 0$ we can again perform the integration in (4.7) via integrating around the contour shown in Fig. (4.2) now with $R = R_N = N\pi$ for $N = 1, 2, \dots$. The integrand is analytic inside $C_\delta^{R_N} \cup C_{R_N}$, except for simple poles at $k = k_n = -i\left(n - \frac{1}{2}\right)\pi$ for $n = 1, 2, \dots, N$. The Cauchy Residue Theorem gives

$$\int_{C_\delta^{R_N}} f(k,x)\, dk + \int_{C_{R_N}} f(k,x)\, dk = -2\pi i \sum_{n=1}^N \text{Res}\,(f(k,x); k_n),$$

where

$$f(k,x) = \frac{\tanh k}{k}\exp(-ikx), \quad (k,x) \in \mathbb{C}\setminus\left\{\pm i\left(n-\frac{1}{2}\right)\pi, n \in \mathbb{N}\right\}\times\mathbb{R}^{+}.$$

We have

$$\mathrm{Res}(f(k,x);k_n) = \frac{2i\exp(-(n-\frac{1}{2})\pi x)}{(2n-1)\pi}$$

and so,

$$\int_{C_\delta^{R_N}} f(k,x)\,dk + \int_{C_{R_N}} f(k,x)\,dk = 4\sum_{n=1}^{N}\frac{\exp(-(n-\frac{1}{2})\pi x)}{2n-1}. \tag{4.10}$$

On C_{R_N}, we again write

$$k = R_N\cos\theta + iR_N\sin\theta \quad \text{for } \theta \in [-\pi,0],$$

so that the integral along C_{R_N} becomes

$$\int_{C_{R_N}} f(k,x)\,dk = -\int_{\theta=-\pi}^{0} f(R(\cos\theta + i\sin\theta),x)R(-\sin\theta + i\cos\theta)\,d\theta.$$

Taking the absolute value and observing that $|\tanh k| \leq 2$ for $k \in C_{R_N}$ gives

$$\left|\int_{C_{R_N}} f(k,x)\,dk\right| \leq \int_{\theta=-\pi}^{0} 2\exp(xR_N\sin\theta)\,d\theta, \quad x > 0, \quad N \in \mathbb{N}. \tag{4.11}$$

We have that $\sin\theta \leq h(\theta)$ for $\theta \in [-\pi,0]$ where

$$h(\theta) = \begin{cases} \dfrac{\theta}{\pi}, & \theta \in \left[-\dfrac{\pi}{2},0\right], \\[2mm] -\dfrac{\theta}{\pi} - 1, & \theta \in \left[-\pi,-\dfrac{\pi}{2}\right], \end{cases}$$

so that

$$\int_{\theta=-\pi}^{0} 2\exp(xR_N\sin\theta)\,d\theta \leq \int_{\theta=-\pi}^{-\frac{\pi}{2}} 2\exp\left(xR_N\left(-\frac{\theta}{\pi}-1\right)\right)d\theta$$
$$+ \int_{\theta=-\frac{\pi}{2}}^{0} 2\exp\left(xR_N\frac{\theta}{\pi}\right)d\theta \tag{4.12}$$

with $x \in \mathbb{R}^{+}, N \in \mathbb{N}$. It follows from (4.12) that

$$\int_{\theta=-\pi}^{0} 2\exp(xR_N\sin\theta)\,d\theta \leq \frac{4\pi}{xR_N}, \quad x \in \mathbb{R}^{+}, N \in \mathbb{N}. \tag{4.13}$$

Hence, via (4.11) and (4.13),

$$\int_{C_{R_N}} f(k,x)\,dk \to 0 \quad \text{as } N \to \infty \text{ with } x \in \mathbb{R}^{+}.$$

Thus from (4.10) we have

$$\int_{C_\delta} f(k,x)\,dk = 4\sum_{n=1}^{\infty} \frac{\exp(-(n-\frac{1}{2})\pi x)}{(2n-1)}, \quad x \in \mathbb{R}^+,$$

and so

$$a_1(x) = -2\sum_{n=1}^{\infty} \frac{\exp(-(n-\frac{1}{2})\pi x)}{(2n-1)}, \quad x \in \mathbb{R}^+.$$

It is now instructive to consider the sum

$$\sum_{n=1}^{\infty} \frac{1}{(2n-1)}\exp\left(-\left(n-\frac{1}{2}\right)\pi x\right) = \sum_{r=1}^{\infty} \frac{1}{r}\exp\left(-\frac{\pi}{2}rx\right) - \sum_{r=1}^{\infty} \frac{1}{2r}\exp\left(-\pi rx\right)$$

$$= S_1(x) - \frac{1}{2}S_2(x), \quad x \in \mathbb{R}^+.$$

Both $S_1(x)$ and $S_2(x)$ define continuous and continuously differentiable functions for $x \in \mathbb{R}^+$, and may be differentiated term by term. Hence

$$S_1'(x) = -\frac{\pi}{2}\sum_{r=1}^{\infty} \exp\left(-\frac{\pi}{2}x\right)^r, \qquad S_2'(x) = -\pi\sum_{r=1}^{\infty} \exp\left(-\pi x\right)^r, \quad x \in \mathbb{R}^+.$$

However both are geometric series, so can be written in closed form as

$$S_1'(x) = -\frac{\pi\exp(-\frac{\pi}{2}x)}{2(1-\exp(-\frac{\pi}{2}x))}, \qquad S_2'(x) = -\frac{\pi\exp(-\pi x)}{1-\exp(-\pi x)}, \quad x \in \mathbb{R}^+,$$

which can be integrated to give

$$S_1(x) = -\log\left(1-\exp\left(-\frac{\pi}{2}x\right)\right) + C_1, \qquad S_2(x) = -\log\left(1-\exp\left(-\pi x\right)\right) + C_2,$$

for $x \in \mathbb{R}^+$, with constants $C_1, C_2 \in \mathbb{R}$. However, $S_1(x), S_2(x) \to 0$ as $x \to \infty$, which requires that $C_1 = C_2 = 0$. Hence we have

$$S_1(x) = -\log\left(1-\exp\left(-\frac{\pi}{2}x\right)\right), \qquad S_2(x) = -\log\left(1-\exp\left(-\pi x\right)\right), \quad x \in \mathbb{R}^+.$$

This leads to the result

$$\sum_{n=1}^{\infty} \frac{1}{(2n-1)}\exp\left(-\left(n-\frac{1}{2}\right)\pi x\right) = \frac{1}{2}\log\left(\frac{(1-\exp(-\pi x))}{\left(1-\exp\left(-\frac{\pi}{2}x\right)\right)^2}\right), \quad x \in \mathbb{R}^+,$$

and so we have

$$a_1(x) = \log\left(\tanh\frac{\pi}{4}x\right), \quad x \in \mathbb{R}^+. \tag{4.14}$$

It follows that,

$$a_1(x) = \begin{cases} \log\dfrac{\pi}{4}x + O(x^2) & \text{as } x \to 0, \\[2mm] -2\exp\left(-\dfrac{\pi}{2}x\right) + O\left(\exp\left(-\dfrac{3\pi}{2}x\right)\right) & \text{as } x \to \infty. \end{cases} \tag{4.15}$$

We now have, via (4.5), (4.9) and (4.14), that

$$I(x,t) = t^2 \log\left(\tanh\frac{\pi}{4}x\right) + o(t^2) \qquad (4.16)$$

as $t \to 0$, uniformly for $x \in \bar{\mathbb{R}}^+\setminus[0,\Delta(t))$. In particular, via (4.15) and (4.16), we have

$$I(x,t) = t^2\left(\log\frac{\pi}{4}x + O(x^2)\right) + o(t^2)$$

as $t \to 0$ with $t^2 \ll x \ll 1$ (recalling that $\Delta(t) = O\left(t^2\right)$ as $t \to 0^+$) whilst,

$$I(x,t) = t^2\left(-2\exp\left(-\frac{\pi}{2}x\right) + O\left(\exp\left(-\frac{3\pi}{2}x\right)\right)\right) + o(t^2) \qquad (4.17)$$

as $t \to 0$ with $x \gg 1$.

Now from (3.32) and (3.34) we may write

$$\bar{\eta}(x,t) = \begin{cases} \dfrac{1}{2\pi\beta}\left(I(x,t) - I(x-\beta,t)\right), & (x,t) \in [\beta,\infty) \times \mathbb{R}^+, \\[2ex] \dfrac{1}{2\pi\beta}\left(2\pi(\beta-x) + I(x,t) - I(\beta-x,t)\right), & (x,t) \in (0,\beta) \times \mathbb{R}^+, \\[2ex] \dfrac{1}{2\pi\beta}\left(2\pi\beta + I(-x,t) - I(\beta-x,t)\right), & (x,t) \in (-\infty,0] \times \mathbb{R}^+. \end{cases}$$

It follows from (4.16) that,

$$\bar{\eta}(x,t) = \begin{cases} \dfrac{t^2}{2\pi\beta}\log\left(\dfrac{\tanh\frac{\pi}{4}x}{\tanh\frac{\pi}{4}(x-\beta)}\right) + o(t^2), & \text{as } t \to 0 \text{ with} \\ & \qquad x \in [\beta+\Delta(t),\infty), \\[2ex] \dfrac{1}{\beta}(\beta-x) + \dfrac{t^2}{2\pi\beta}\log\left(\dfrac{\tanh\frac{\pi}{4}x}{\tanh\frac{\pi}{4}(\beta-x)}\right) + o(t^2), & \text{as } t \to 0 \text{ with} \\ & \qquad x \in [\Delta(t),\beta-\Delta(t)], \\[2ex] 1 + \dfrac{t^2}{2\pi\beta}\log\left(\dfrac{\tanh\frac{\pi}{4}(-x)}{\tanh\frac{\pi}{4}(\beta-x)}\right) + o(t^2), & \text{as } t \to 0 \text{ with} \\ & \qquad x \in (-\infty,-\Delta(t)]. \end{cases}$$

$$(4.18)$$

It is instructive to examine the function

$$F(x) = \log\left(\left|\frac{\tanh\frac{\pi}{4}x}{\tanh\frac{\pi}{4}(x-\beta)}\right|\right), \quad x \in \mathbb{R}\setminus\{0,\beta\}.$$

As expected, $F \in C^1\left(\mathbb{R}\backslash\{0,\beta\}\right)$. In addition we have

$$
F(x) = \begin{cases}
-\log(x-\beta) + \log\left(\dfrac{4}{\pi}\tanh\dfrac{\pi}{4}\beta\right) + O((x-\beta)^2) & \text{as } x \to \beta^+, \\[3mm]
-\log(\beta-x) + \log\left(\dfrac{4}{\pi}\tanh\dfrac{\pi}{4}\beta\right) + O((x-\beta)^2) & \text{as } x \to \beta^-,
\end{cases}
$$

whilst

$$
F(x) = \begin{cases}
\log x - \log\left(\dfrac{4}{\pi}\tanh\dfrac{\pi}{4}\beta\right) + O(x^2) & \text{as } x \to 0^+, \\[3mm]
\log(-x) - \log\left(\dfrac{4}{\pi}\tanh\dfrac{\pi}{4}\beta\right) + O(x^2) & \text{as } x \to 0^-,
\end{cases}
$$

and

$$
F(x) = \begin{cases}
2\left(\exp\left(\dfrac{\pi}{2}\beta\right)-1\right)\exp\left(-\dfrac{\pi}{2}x\right) + O\left(\exp\left(-\dfrac{3\pi}{2}x\right)\right) & \text{as } x \to \infty, \\[3mm]
2\left(\exp\left(-\dfrac{\pi}{2}\beta\right)-1\right)\exp\left(\dfrac{\pi}{2}x\right) + O\left(\exp\left(\dfrac{3\pi}{2}x\right)\right) & \text{as } x \to -\infty.
\end{cases}
$$

Also, $F(x)$ has exactly one zero, at $x = \frac{1}{2}\beta$, and $F(x)$ is positive and monotone decreasing in $x > \beta$, monotone increasing in $0 < x < \beta$, and negative and monotone decreasing in $x < 0$. A graph of $F(x)$ for $\beta = 1$ is given in Fig. (4.3). We can now write,

$$
\bar{\eta}(x,t) = \eta_0(x) + \frac{t^2}{2\pi\beta}\log\left(\left|\frac{\tanh\frac{\pi}{4}x}{\tanh\frac{\pi}{4}(x-\beta)}\right|\right) + o(t^2), \quad \text{as } t \to 0
$$

with $x \in (-\infty, -\Delta(t)] \cup [\Delta(t), \beta - \Delta(t)] \cup [\beta + \Delta(t), \infty)$, and $\Delta(t) = O(t^2)$ as $t \to 0$.

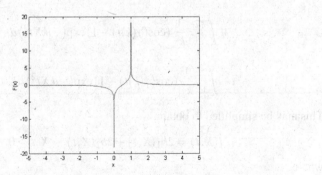

Fig. 4.3: The graph of $F(x)$ with $\beta = 1$.

4.2   Inner Region Coordinate Expansion for $I(x,t)$ as $t \to 0$

We now examine $I(x,t)$ for $x \in [0, \Delta(t))$. We write $x = Xt^2$ with $X \geq 0$ and $X = O(1)$ as $t \to 0$ and write $I(Xt^2, t) = \hat{I}(X,t)$. The integral (3.33) then becomes

$$\hat{I}(X,t) = \int_{C_\delta} \frac{1}{k^2} \cos(\gamma(k)t) \exp(-ikXt^2)\, dk, \quad X, t \geq 0.$$

We now write this as

$$\hat{I}(X,t) = \int_{C_\delta} \frac{1}{k^2} \left(\cos(\gamma(k)t) - 1 \right) \exp(-ikXt^2)\, dk + \int_{C_\delta} \frac{1}{k^2} \exp(-it^2 kX)\, dk,$$

for $X, t \geq 0$. Via (4.9) we have that

$$\int_{C_\delta} \frac{1}{k^2} \exp(-it^2 kX)\, dk = 0, \quad X, t \geq 0,$$

and so

$$\hat{I}(X,t) = \int_{C_\delta} \frac{1}{k^2} \left(\cos(\gamma(k)t) - 1 \right) \exp(-ikXt^2)\, dk \quad X, t \geq 0.$$

The singularity of the integrand at $k = 0$ is removable, and so, via the Cauchy Residue Theorem, we may deform the contour C_δ onto the real k-axis. Thus

$$\hat{I}(X,t) = \int_{-\infty}^{\infty} \frac{1}{k^2} \left(\cos(\gamma(k)t) - 1 \right) \exp(-ikXt^2)\, dk \quad X, t \geq 0. \tag{4.19}$$

The integrand may be expanded about $t = 0$ in an asymptotic form provided that $|k| \ll \frac{1}{t^2}$. Thus we introduce $\delta(t)$ such that $t^2 \ll \delta(t) \ll 1$ as $t \to 0$, after which we write (4.19) as

$$\hat{I}(X,t) = \int_{-\infty}^{-\frac{1}{\delta(t)}} \frac{1}{k^2} \left(\cos(\gamma(k)t) - 1 \right) \exp(-ikXt^2)\, dk$$

$$+ \int_{-\frac{1}{\delta(t)}}^{\frac{1}{\delta(t)}} \frac{1}{k^2} \left(\cos(\gamma(k)t) - 1 \right) \exp(-ikXt^2)\, dk$$

$$+ \int_{\frac{1}{\delta(t)}}^{\infty} \frac{1}{k^2} \left(\cos(\gamma(k)t) - 1 \right) \exp(-ikXt^2)\, dk, \quad X, t \geq 0.$$

This may be simplified to obtain,

$$\hat{I}(X,t) = 2b_1(X,t) + 2b_2(X,t), \quad X, t \geq 0, \tag{4.20}$$

where

$$b_1(X,t) = \int_0^{\frac{1}{\delta(t)}} \frac{1}{k^2} \left(\cos(\gamma(k)t) - 1 \right) \cos(kXt^2)\, dk, \quad X, t \geq 0, \tag{4.21}$$

$$b_2(X,t) = \int_{\frac{1}{\delta(t)}}^{\infty} \frac{1}{k^2} \left(\cos(\gamma(k)t) - 1\right) \cos(kXt^2) \, dk, \quad X, t \geq 0. \tag{4.22}$$

We first consider $b_2(x,t)$ as $t \to 0$ with $X(\geq 0) = O(1)$. We observe that,

$$\gamma(k) = k^{\frac{1}{2}} + O\left(k^{\frac{1}{2}} \exp(-2k)\right), \quad \text{as } k \to \infty.$$

Now, in (4.22), $k \geq \delta(t)^{-1} \gg 1$ as $t \to 0$. Thus, over the range of integration in (4.22), we have,

$$\cos(\gamma(k)t) - 1 = \left(\cos\left(k^{\frac{1}{2}}t\right) - 1\right) + O\left(\frac{t}{\delta(t)^{\frac{1}{2}}} \exp\left(-\frac{2}{\delta(t)}\right)\right)$$

as $t \to 0$ uniformly for $k \in [\delta(t)^{-1}, \infty)$. Therefore we have,

$$b_2(X,t) = \int_{\frac{1}{\delta(t)}}^{\infty} \frac{1}{k^2} \left(\cos(k^{\frac{1}{2}}t) - 1\right) \cos(kXt^2) \, dk + O\left(t\delta(t)^{\frac{1}{2}} \exp\left(-\frac{2}{\delta(t)}\right)\right), \tag{4.23}$$

as $t \to 0$ with $X(\geq 0) = O(1)$. We next make the substitution $s = t^2 k$ in the integral in (4.23), which becomes

$$\int_{\frac{1}{\delta(t)}}^{\infty} \frac{1}{k^2} \left(\cos(k^{\frac{1}{2}}t) - 1\right) \cos(kXt^2) \, dk = t^2 \int_{\frac{t^2}{\delta(t)}}^{\infty} \frac{1}{s^2} \left(\cos(s^{\frac{1}{2}}) - 1\right) \cos(Xs) \, ds, \tag{4.24}$$

for $X \geq 0$. We recall that $t^2 \delta(t)^{-1} \ll 1$ as $t \to 0$, whilst

$$\frac{\cos(s^{\frac{1}{2}}) - 1}{s^2} = -\frac{1}{2s} + \frac{1}{24} + O(s), \quad \text{as } s \to 0. \tag{4.25}$$

With this in mind, we write,

$$\int_{\frac{t^2}{\delta(t)}}^{\infty} \frac{1}{s^2} \left(\cos(s^{\frac{1}{2}}) - 1\right) \cos(Xs) \, ds = \int_{\frac{t^2}{\delta(t)}}^{1} \frac{1}{s^2} \left(\cos(s^{\frac{1}{2}}) - 1\right) \cos(Xs) \, ds$$

$$+ \int_{1}^{\infty} \frac{1}{s^2} \left(\cos(s^{\frac{1}{2}}) - 1\right) \cos(Xs) \, ds. \tag{4.26}$$

We now write,

$$\int_{\frac{t^2}{\delta(t)}}^{1} \frac{1}{s^2} \left(\cos(s^{\frac{1}{2}}) - 1\right) \cos(Xs) \, ds$$

$$= \int_{\frac{t^2}{\delta(t)}}^{1} \left(\frac{1}{s^2} \left(\cos(s^{\frac{1}{2}}) - 1\right) \cos(Xs) + \frac{1}{2s}\right) ds - \int_{\frac{t^2}{\delta(t)}}^{1} \frac{1}{2s} \, ds \tag{4.27}$$

$$= \int_{\frac{t^2}{\delta(t)}}^{1} \left(\frac{1}{s^2} \left(\cos(s^{\frac{1}{2}}) - 1\right) \cos(Xs) + \frac{1}{2s}\right) ds + \log\left(\frac{t}{\delta(t)^{\frac{1}{2}}}\right).$$

We observe that

$$\int_{\frac{t^2}{\delta(t)}}^{1} \left(\frac{1}{s^2} \left(\cos(s^{\frac{1}{2}}) - 1 \right) \cos(Xs) + \frac{1}{2s} \right) ds = \int_{\frac{t^2}{\delta(t)}}^{1} h(s) \cos(Xs) \, ds$$

$$+ \int_{\frac{t^2}{\delta(t)}}^{1} \frac{1}{2s} \left(1 - \cos(Xs) \right) ds,$$

(4.28)

where

$$h(s) = \begin{cases} \dfrac{1}{2s} \left(1 + \dfrac{2}{s} \left(\cos\left(s^{\frac{1}{2}} \right) - 1 \right) \right), & s > 0, \\[2mm] \dfrac{1}{24}, & s = 0. \end{cases}$$

(4.29)

and we note that $h \in C^1(\bar{\mathbb{R}}^+)$. It now follows from (4.28) that,

$$\int_{\frac{t^2}{\delta(t)}}^{1} \left(\frac{1}{s^2} \left(\cos(s^{\frac{1}{2}}) - 1 \right) \cos(Xs) + \frac{1}{2s} \right) ds$$

(4.30)

$$= \left\{ \int_0^1 h(s) \cos(Xs) \, ds + \int_0^1 \frac{1}{2s} \left(1 - \cos(Xs) \right) ds. \right\} + O\left(\frac{t^2}{\delta(t)} \right)$$

as $t \to 0$ with $X(\geq 0) = O(1)$. Thus, via (4.24)–(4.30), we have,

$$\int_{\frac{1}{\delta(t)}}^{\infty} \frac{1}{k^2} \left(\cos(k^{\frac{1}{2}}t) - 1 \right) \cos(kXt^2) \, dk$$

$$= t^2 \left[\log \left(\frac{t}{\delta(t)^{\frac{1}{2}}} \right) + (F_1(X) + F_2(X) + F_3(X)) + O\left(\frac{t^2}{\delta(t)} \right) \right],$$

(4.31)

as $t \to 0$ with $X(\geq 0) = O(1)$, where $F_1, F_2, F_3 \in C^1(\bar{\mathbb{R}}^+)$, and

$$F_1(X) = \int_1^{\infty} \frac{1}{s^2} \left(\cos(s^{\frac{1}{2}}) - 1 \right) \cos(Xs) \, ds,$$

(4.32)

$$F_2(X) = \int_0^1 \frac{1}{2s} \left(1 - \cos(Xs) \right) ds,$$

(4.33)

$$F_3(X) = \int_0^1 h(s) \cos(Xs) \, ds,$$

(4.34)

in $X \geq 0$. It should be noted that we may re-write $F_2(X)$ in the form

$$F_2(X) = \int_0^X \frac{1}{2u} \left(1 - \cos(u) \right) du,$$

(4.35)

in $X \geq 0$, which will be of use at a later stage.

Next we consider the forms of $F_1(X)$, $F_2(X)$ and $F_3(X)$ as $X \to 0^+$ and $X \to \infty$ respectively. We address $F_1(X)$ first. It is convenient to write

$$F_1(X) = \int_1^\infty \frac{1}{s^2} \cos(s^{\frac{1}{2}}) \cos(Xs) \, ds - \int_1^\infty \frac{1}{s^2} \cos(Xs) \, ds \qquad (4.36)$$

in $X \geq 0$. We now consider the limit $X \to 0^+$ in (4.36). Beginning with the second integral in (4.36), we write

$$\int_1^\infty \frac{1}{s^2} \cos(Xs) \, ds = X \int_X^\infty \frac{1}{u^2} \cos u \, du$$

$$= X \left(\int_X^\infty \frac{1}{u^2} (\cos u - 1) \, du + \int_X^\infty \frac{1}{u^2} \, du \right)$$

$$= 1 - \left(\int_0^\infty \frac{1 - \cos u}{u^2} \, du \right) X + O(X^2) \quad \text{as } X \to 0^+.$$

However, it is readily established, via contour integration, that

$$\int_0^\infty \frac{1 - \cos u}{u^2} \, du = \frac{1}{2}\pi.$$

Therefore,

$$\int_1^\infty \frac{1}{s^2} \cos(Xs) \, ds = 1 - \frac{1}{2}\pi X + O(X^2) \quad \text{as } X \to 0^+. \qquad (4.37)$$

Next we consider the first integral in (4.36). We write

$$\int_1^\infty \frac{1}{s^2} \cos(s^{\frac{1}{2}}) \cos(Xs) \, ds = \int_1^\infty \frac{1}{s^2} \cos(s^{\frac{1}{2}}(\cos(Xs) - 1) \, ds + \int_1^\infty \frac{1}{s^2} \cos(s^{\frac{1}{2}}) \, ds$$

$$= X \int_X^\infty \frac{1}{u^2} \cos\left(\frac{1}{\sqrt{X}} u^{\frac{1}{2}}\right) (\cos u - 1) \, du + c_1, \qquad (4.38)$$

where we have set $u = Xs$, with

$$c_1 = \int_1^\infty \frac{1}{s^2} \cos(s^{\frac{1}{2}}) \, ds = 0.036242\ldots.$$

Now,

$$\int_X^\infty \frac{1}{u^2} \cos\left(\frac{1}{\sqrt{X}} u^{\frac{1}{2}}\right) (\cos u - 1) \, du = \int_0^\infty \frac{1}{u^2} \cos\left(\frac{1}{\sqrt{X}} u^{\frac{1}{2}}\right) (\cos u - 1) \, du$$

$$- \int_0^X \frac{1}{u^2} \cos\left(\frac{1}{\sqrt{X}} u^{\frac{1}{2}}\right) (\cos u - 1) \, du$$

$$= \int_0^\infty \frac{1}{u^2} \cos\left(\frac{1}{\sqrt{X}} u^{\frac{1}{2}}\right) (\cos u - 1) \, du + O(X) \qquad (4.39)$$

as $X \to 0^+$. In addition, an integration by parts establishes that,

$$\int_0^\infty \frac{1}{u^2} \cos\left(\frac{1}{\sqrt{X}} u^{\frac{1}{2}}\right) (\cos u - 1)\, du = o(\sqrt{X}), \quad \text{as } X \to 0^+. \tag{4.40}$$

It then follows from (4.39) and (4.40) that

$$\int_X^\infty \frac{1}{u^2} \cos\left(\frac{1}{\sqrt{X}} u^{\frac{1}{2}}\right) (\cos u - 1)\, du = o(\sqrt{X}), \quad \text{as } X \to 0^+. \tag{4.41}$$

Therefore, via (4.38) and (4.41),

$$\int_1^\infty \frac{1}{s^2} \cos(s^{\frac{1}{2}}) \cos(Xs)\, ds = c_1 + o(X^{\frac{3}{2}}), \quad \text{as } X \to 0^+. \tag{4.42}$$

Finally, we have, via (4.36), (4.37) and (4.42), that

$$F_1(X) = (c_1 - 1) + \frac{1}{2}\pi X + o\left(X^{\frac{3}{2}}\right) \quad \text{as } X \to 0^+. \tag{4.43}$$

Next we consider $F_1(X)$, given in (4.32), as $X \to \infty$. It is convenient to introduce

$$g(s) = \frac{1}{s^2}\left(\cos(s^{\frac{1}{2}}) - 1\right), \quad s \geq 1, \tag{4.44}$$

and observe that $g \in C^\infty([1, \infty))$. Now, via (4.32) and (4.44) we write

$$F_1(X) = \int_1^\infty g(s) \cos(Xs)\, ds$$

$$= \left[g(s) \frac{\sin(Xs)}{X}\right]_1^\infty - \frac{1}{X}\int_1^\infty g'(s) \sin(Xs)\, ds$$

$$= -g(1)\frac{\sin X}{X} - \frac{1}{X}\left[-g'(s)\frac{\cos(Xs)}{X}\right]_1^\infty + \frac{1}{X^2}\int_1^\infty g''(s) \cos(Xs)\, ds$$

$$= -g(1)\frac{\sin X}{X} - g'(1)\frac{\cos X}{X^2} + o\left(\frac{1}{X^2}\right) \quad \text{as } X \to \infty,$$

via the Riemann-Lebesgue Lemma. From (4.44) we have,

$$g(1) = \cos(1) - 1$$

$$g'(1) = -\frac{1}{2}\sin(1) - 2(\cos(1) - 1)$$

Hence,

$$F_1(X) = (1 - \cos(1))\frac{\sin X}{X} + \left(\frac{1}{2}\sin(1) - 2(1 - \cos(1))\right)\frac{\cos X}{X^2} + o\left(\frac{1}{X^2}\right), \tag{4.45}$$

as $X \to \infty$. We now address $F_2(X)$, given in (4.35). We begin by examining $F_2(X)$ as $X \to 0^+$. It follows from (4.35) that

$$F_2(X) = \frac{1}{2} \int_0^X \left(\frac{1}{2}u - \frac{1}{24}u^3 + O(u^5) \right) du, \quad X \geq 0.$$

Hence,

$$F_2(X) = \frac{1}{8}X^2 + O(X^4), \quad \text{as } X \to 0^+. \tag{4.46}$$

Next we consider $F_2(X)$ as $X \to \infty$. We write,

$$F_2(X) = \int_0^1 \frac{1 - \cos u}{2u} du + \int_1^X \frac{1 - \cos u}{2u} du$$

$$= \int_0^1 \frac{1 - \cos u}{2u} du + \frac{1}{2} \log X - \frac{1}{2} \left\{ \int_1^\infty \frac{\cos u}{u} du - \int_X^\infty \frac{\cos u}{u} du \right\}, \quad X \geq 0.$$

Hence,

$$F_2(X) = \frac{1}{2} \log X + c_2 - \frac{\sin X}{2X} + \frac{\cos X}{2X^2} + O\left(\frac{1}{X^3} \right), \quad \text{as } X \to \infty \tag{4.47}$$

with

$$c_2 = \int_0^1 \frac{1 - \cos u}{2u} du - \int_1^\infty \frac{\cos u}{2u} du = 0.288608\dots. \tag{4.48}$$

We now address $F_3(X)$, given in (4.34), as $X \to 0^+$. It follows from (4.34) that

$$F_3(X) = \int_0^1 h(s) \left(1 - \frac{1}{2}X^2 s^2 + O(X^4 s^4) \right) ds \quad \text{as } X \to 0^+.$$

Hence

$$F_3(X) = c_3 - d_3 X^2 + O(X^4) \quad \text{as } X \to 0^+, \tag{4.49}$$

with

$$c_3 = \int_0^1 h(s) \, ds = 0.040980\dots$$

and

$$d_3 = \frac{1}{2} \int_0^1 h(s) s^2 \, ds = 0.006773\dots.$$

We now address $F_3(X)$ as $X \to \infty$. We write, via integration by parts in (4.34),

$$F_3(X) = \left[h(s) \frac{\sin(Xs)}{X} \right]_0^1 - \frac{1}{X} \int_0^1 h'(s) \sin(Xs) \, ds$$

$$= h(1) \frac{\sin X}{X} - \frac{1}{X^2} \left[-h'(s) \cos(Xs) \right]_0^1 - \frac{1}{X^2} \int_0^1 h''(s) \cos(Xs) \, ds$$

$$= h(1) \frac{\sin X}{X} + \frac{1}{X^2} \left(h'(1) \cos(X) - h'(0) \right) + o\left(\frac{1}{X^2} \right) \quad \text{as } X \to \infty$$

via the Riemann-Lebesgue Lemma. From (4.29) we have,

$$h(1) = \cos(1) - \frac{1}{2},$$

$$h'(1) = \frac{3}{2} - \frac{1}{2}\sin(1) - 2\cos(1),$$

$$h'(0) = -\frac{1}{720}.$$

Hence

$$F_3(X) = \left(\cos(1) - \frac{1}{2}\right)\frac{\sin X}{X} + \frac{1}{X^2}\left(\left(\frac{3}{2} - \frac{1}{2}\sin(1) - 2\cos(1)\right)\cos(X) + \frac{1}{720}\right)$$
$$+ o\left(\frac{1}{X^2}\right)$$

(4.50)

as $X \to \infty$. We now have, via (4.22), (4.23) and (4.31), that

$$b_2(X,t) = t^2\left(\log\left(\frac{t}{\delta(t)^{\frac{1}{2}}}\right) + (F_1(X) + F_2(X) + F_3(X))\right) + o(t^2) \quad (4.51)$$

as $t \to 0$ with $X(\geq 0) = O(1)$.

We now turn our attention to $b_1(X,t)$, given in (4.21), as $t \to 0$ with $X(\geq 0) = O(1)$. Since $t^2 \ll \delta(t) \ll 1$ as $t \to 0$, over the range of integration in (4.21)

$$\cos\left(kXt^2\right) = 1 + O\left(t^4\delta(t)^{-2}\right) \quad \text{as } t \to 0 \text{ with } X(\geq 0) = O(1).$$

In addition, we have, over the range of integration,

$$\cos\gamma(k)t = 1 - \frac{1}{2}\gamma^2(k)t^2 + O\left(t^4\delta(t)^{-2}\right) \quad \text{as } t \to 0.$$

Thus we have,

$$b_1(X,t) = -\frac{1}{2}t^2\int_0^{\frac{1}{\delta(t)}}\frac{\tanh k}{k}\,dk + o(t^2)$$

as $t \to 0$ with $X(\geq 0) = O(1)$. We now write

$$\int_0^{\frac{1}{\delta(t)}}\frac{\tanh k}{k}\,dk = \int_0^1\frac{\tanh k}{k}\,dk + \int_1^{\frac{1}{\delta(t)}}\frac{\tanh k}{k}\,dk$$

$$= \int_0^1\frac{\tanh k}{k}\,dk + \int_1^{\frac{1}{\delta(t)}}\frac{\tanh k - 1}{k}\,dk + \int_1^{\frac{1}{\delta(t)}}\frac{1}{k}\,dk$$

$$= \int_0^1\frac{\tanh k}{k}\,dk + \int_1^{\frac{1}{\delta(t)}}\frac{\tanh k - 1}{k}\,dk + \log\frac{1}{\delta(t)}.$$

Thus,

$$b_1(X,t) = -\frac{1}{2}t^2 \left(\log\left(\frac{1}{\delta(t)}\right) + c_4 \right) + o(t^2), \tag{4.52}$$

as $t \to 0$ with $X(\geq 0) = O(1)$, and,

$$c_4 = \int_0^1 \frac{\tanh k}{k}\,dk + \int_1^\infty \frac{\tanh k - 1}{k}\,dk = 0.818780\ldots. \tag{4.53}$$

Finally, via (4.20), (4.21), (4.22), (4.51) and (4.52), we have

$$\hat{I}(X,t) = 2t^2 \left(\log t + F_1(X) + F_2(X) + F_3(X) - \frac{1}{2}c_4 \right) + o(t^2)$$

as $t \to 0$ with $X(\geq 0) = O(1)$. We note, that when $0 \leq X \ll 1$, then, via (4.43), (4.46) and (4.49), we have

$$\hat{I}(X,t) = 2t^2 \left(\log t + \left(c_1 + c_3 - \frac{1}{2}c_4 - 1 \right) + \frac{1}{2}\pi X + o\left(X^{\frac{3}{2}}\right) \right) + o(t^2) \text{ as } t \to 0,$$

whilst, when $X \gg 1$, via (4.45), (4.47) and (4.50), we have,

$$\hat{I}(X,t) = 2t^2 \left(\log t + \frac{1}{2}\log X + \left(c_2 - \frac{1}{2}c_4 \right) + O\left(\frac{1}{X^2}\right) \right) + o(t^2) \quad \text{as } t \to 0.$$

In addition, it follows from (3.34) that,

$$\hat{I}(X,t) = 2t^2 \left(\log t + F_1(-X) + F_2(-X) + F_3(-X) + \pi X - \frac{1}{2}c_4 \right) + o(t^2) \tag{4.54}$$

as $t \to 0$ with $X(\leq 0) = O(1)$. It should be noted, via (4.43), (4.46), (4.49), and (4.54) that $\hat{I}(X,t)$ is continuous with continuous derivative $\hat{I}_X(X,t)$ for $-\infty < X < \infty$. In particular, when $0 \leq (-X) \ll 1$ we have, via (4.43), (4.46), (4.49), and (4.54),

$$\hat{I}(X,t) = 2t^2 \left(\log t + \left(c_1 + c_3 - \frac{1}{2}c_4 - 1 \right) + \frac{1}{2}\pi X + o\left((-X)^{\frac{3}{2}}\right) \right) \quad \text{as } t \to 0,$$

whilst, when $(-X) \gg 1$ we have, via (4.45), (4.47), (4.50), and (4.54),

$$\hat{I}(X,t) = 2t^2 \left(\log t + \pi X + \frac{1}{2}\log(-X) + \left(c_2 - \frac{1}{2}c_4 \right) + O\left(\frac{1}{X^2}\right) \right) + o(t^2),$$

as $t \to 0$. Graphs of $F_1(X)$, $F_2(X)$ and $F_3(X)$ (computed numerically from (4.32), (4.33) and (4.34) via Simpson's method) are given in Figs. (4.4)–(4.12). It is now convenient to write

$$\hat{I}(X,t) = 2t^2 \left(\log t + H(X) \right) + o(t^2) \tag{4.55}$$

as $t \to 0$ with $X = O(1)$. Here

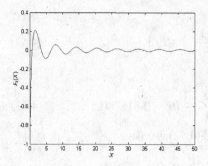

Fig. 4.4: The graph of $F_1(X)$.

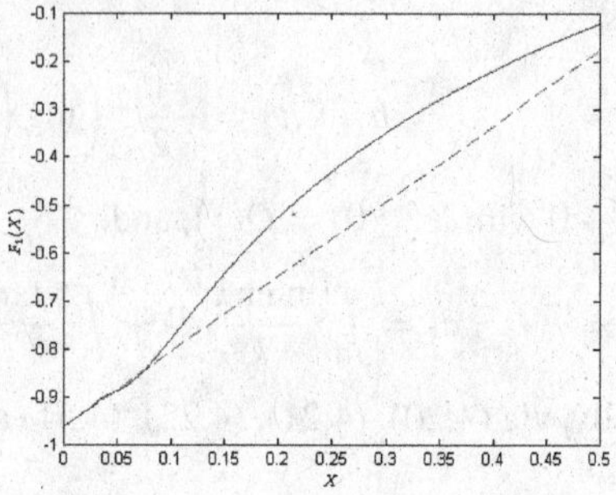

Fig. 4.5: The graph of $F_1(X)$ with asymptotic approximation ($--$) (4.43) for $X \ll 1$.

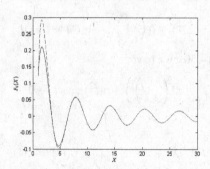

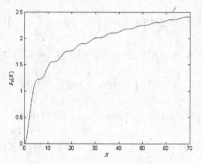

Fig. 4.6: The graph of $F_1(X)$ with asymptotic approximation ($--$) (4.45) for $X \gg 1$.

Fig. 4.7: The graph of $F_2(X)$.

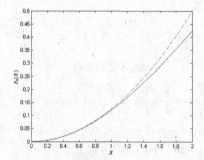

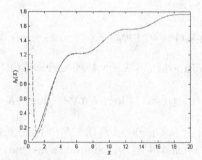

Fig. 4.8: The graph of $F_2(X)$, with asymptotic approximation ($--$) (4.46) for $X \ll 1$.

Fig. 4.9: The graph of $F_2(X)$, with asymptotic approximation ($--$) (4.47) for $X \gg 1$.

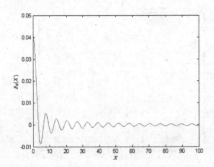 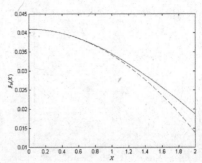

Fig. 4.10: The graph of $F_3(X)$.   Fig. 4.11: The graph of $F_3(X)$ with asymptotic approximation $(--)$ (4.49) for $X \ll 1$.

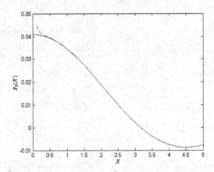

Fig. 4.12: The graph of $F_3(X)$, with asymptotic approximation $(--)$ (4.50) for $X \gg 1$.

$$H(X) = \begin{cases} F_1(X) + F_2(X) + F_3(X) - \dfrac{1}{2}c_4, & X \geq 0 \\[2ex] F_1(-X) + F_2(-X) + F_3(-X) + \pi X - \dfrac{1}{2}c_4, & X < 0, \end{cases} \tag{4.56}$$

and graphs of $H(X)$ for $-\infty < X < \infty$ are given in Figs. (4.13)–(4.16). We note that $H'(0) = \frac{1}{2}\pi$, and, via (4.43), (4.45), (4.46), (4.47), (4.49), (4.50), (4.56) and (4.59), that

$$H(X) = \left(c_1 + c_3 - \frac{1}{2}c_4 - 1\right) + \frac{1}{2}\pi X + o\left(|X|^{\frac{3}{2}}\right)$$

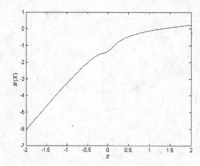

Fig. 4.13: The graph of $H(X)$.

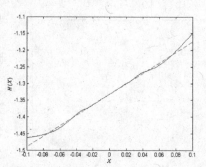

Fig. 4.14: The graph of $H(X)$, with asymptotic approximation ($--$) from (4.43), (4.46), (4.49) and (4.56) for $|X| \ll 1$.

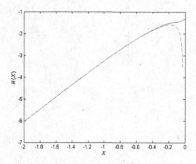

Fig. 4.15: The graph of $H(X)$, with asymptotic approximation ($--$) from (4.45), (4.47), (4.50) and (4.56) for $(-X) \gg 1$.

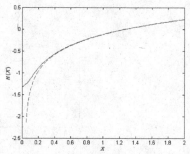

Fig. 4.16: The graph of $H(X)$, with asymptotic approximation ($--$) from (4.45), (4.47), (4.50) and (4.56) for $X \gg 1$.

as $X \to 0$, whilst,

$$H(X) = \begin{cases} \dfrac{1}{2}\log\left(\dfrac{\pi}{4}X\right) + O\left(\dfrac{1}{X^2}\right) & \text{as } X \to +\infty, \\[2mm] \pi X + \dfrac{1}{2}\log\left(-\dfrac{\pi}{4}X\right) + O\left(\dfrac{1}{X^2}\right) & \text{as } X \to -\infty. \end{cases}$$

It is worth recapping here that we have constructed a two region asymptotic expansion for $I(x,t)$ in $x \geq 0$ as $t \to 0$. We have, via (4.16) and (4.55),

Region I, $x(>0) \geq O(1)$ as $t \to 0$,

$$I(x,t) = t^2 \log\left(\tanh\frac{\pi}{4}x\right) + o(t^2) \tag{4.57}$$

as $t \to 0$, with $x(>0) = O(1)$.

Region II, $x(\geq 0) = O(t^2)$ as $t \to 0$.

$$I(x,t) = \hat{I}(X,t) = 2t^2\left(\log t + H(X)\right) + o(t^2) \tag{4.58}$$

as $t \to 0$, with $X = \frac{x}{t^2}(\geq 0) = O(1)$.

The structure of $I(x,t)$ in $x \leq 0$ as $t \to 0$ follows from (3.34). We note that the asymptotic expansion (4.57) in region I (as $x \to 0$) and the asymptotic expansion (4.58) in region II (as $X \to \infty$) asymptotically match, according to the asymptotic matching principal of Van Dyke [Van Dyke (1975)], provided

$$c_4 = 2c_2 - \log\frac{\pi}{4}, \tag{4.59}$$

which is readily verified numerically (via (4.48) and (4.53)).

4.3 Coordinate Expansion for $\bar{\eta}(x,t)$ as $t \to 0$

We are now in a position to construct the asymptotic form for $\bar{\eta}(x,t)$ as $t \to 0$ when $x = O(t^2)$ and when $x = \beta + O(t^2)$. When $x = \beta + O(t^2)$ we write $x = \beta + \bar{X}t^2$, with $\bar{X} = O(1)$ as $t \to 0$. We then obtain via (3.32), (4.57) and (4.58),

$$\bar{\eta}(\bar{X},t) = \frac{t^2}{\pi\beta}\left(-\log t - H(\bar{X}) + \frac{1}{2}\log\left(\tanh\frac{\pi}{4}\beta\right)\right) + o(t^2) \tag{4.60}$$

as $t \to 0$ with $\bar{X} = O(1)$. Similarly, when $x = O(t^2)$, we write $x = Xt^2$, with $X = O(1)$ as $t \to 0$. We then obtain via (3.32), (3.34), (4.57) and (4.58),

$$\bar{\eta}(X,t) = 1 + \frac{t^2}{\pi\beta}\left(\log t + H(X) - \pi X - \frac{1}{2}\log\left(\tanh\frac{\pi}{4}\beta\right)\right) + o(t^2) \tag{4.61}$$

as $t \to 0$ with $X = O(1)$.

We have now completed the detailed asymptotic structure of $\bar{\eta}(x,t)$ as $t \to 0$. This has involved three distinct asymptotic regions as follows: $x \in \mathbb{R}\setminus\{N_0(t) \cup N_\beta(t)\}$, $x \in N_0(t)$ and $x \in N_\beta(t)$. Here $N_0(t)$ and $N_\beta(t)$ are $O(t^2)$ neighbourhoods of the points $x = 0$ and $x = \beta$ respectively. Finally, via (4.18), (4.32), (4.33), (4.34), (4.56), (4.57), (4.58), (4.60), and (4.61), we have

Inner Region A, $x \in N_0(t)$ as $t \to 0$.

In inner region A,

$$\bar{\eta}(X,t) = 1 + \frac{t^2}{\pi\beta}\left(\log t + H(X) - \pi X - \frac{1}{2}\log\left(\tanh\frac{\pi}{4}\beta\right)\right) + o(t^2) \quad (4.62)$$

for $X = O(1)$ as $t \to 0$, with $x = t^2 X$.

Inner Region B, $x \in N_\beta(t)$ as $t \to 0$.

In inner region B,

$$\bar{\eta}(\bar{X},t) = \frac{t^2}{\pi\beta}\left(-\log t - H(\bar{X}) + \frac{1}{2}\log\left(\tanh\frac{\pi}{4}\beta\right)\right) + o(t^2) \quad (4.63)$$

for $\bar{X} = O(1)$ as $t \to 0$, with $x = \beta + \bar{X}t^2$.

Outer Region, $x \in \mathbb{R}\backslash\{N_0(t) \cup N_\beta(t)\}$ as $t \to 0$.

In the outer region,

$$\bar{\eta}(x,t) = \begin{cases} \dfrac{t^2}{2\pi\beta}\log\left(\dfrac{\tanh\frac{\pi}{4}x}{\tanh\frac{\pi}{4}(x-\beta)}\right) + o(t^2), & \text{as } t \to 0 \text{ with} \\[2mm] & x \in [\beta,\infty)\backslash N_\beta(t), \\[4mm] \dfrac{1}{\beta}(\beta - x) + \dfrac{t^2}{2\pi\beta}\log\left(\dfrac{\tanh\frac{\pi}{4}x}{\tanh\frac{\pi}{4}(\beta - x)}\right) + o(t^2), & \text{as } t \to 0 \text{ with} \\[2mm] & x \in [0,\beta]\backslash\{N_0(t)\cup N_\beta(t)\}, \\[4mm] 1 + \dfrac{t^2}{2\pi\beta}\log\left(\dfrac{\tanh\frac{\pi}{4}(-x)}{\tanh\frac{\pi}{4}(\beta - x)}\right) + o(t^2), & \text{as } t \to 0 \text{ with} \\[2mm] & x \in (-\infty,0]\backslash N_0(t). \end{cases}$$

$$(4.64)$$

An illustration of the asymptotic structure for $\bar{\eta}(x,t)$ as $t \to 0$, with $x \in \mathbb{R}$, is shown in Fig. (4.17). Graphs of $\bar{\eta}(X,t)$ in inner region A and in inner region B, for $t \in [0,0.1]$, are shown in Fig. (4.18). In Figs. (4.17) and (4.18) we note that close to the initial corners at $x = \beta$ and $x = 0$, the structure of $\bar{\eta}(x,t)$ as $t \to 0$ shows incipient localised jet formation close to $x = \beta$ (in inner region B) and incipient localised collapse close to $x = 0$ (in inner region A).

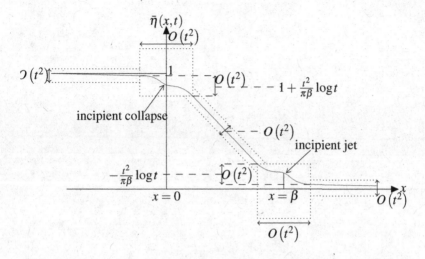

Fig. 4.17: A sketch for the asymptotic structure of $\bar{\eta}(x,t)$ as $t \to 0$.

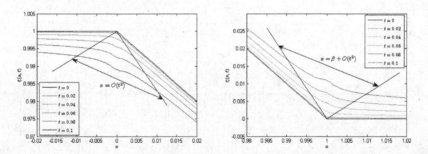

Fig. 4.18: Graphs of $\bar{\eta}(x,t)$ in inner region A and inner region B with $t \in [0, 0.1]$, illustrating the incipient localised collapse and jet structure respectively.

Chapter 5

Coordinate Expansions for $\bar{\eta}(x,t)$ as $|x| \to \infty$

In this chapter we consider $\bar{\eta}(x,t)$, as given by (3.30), in the far fields as $|x| \to \infty$ for $t = O(1)$. We begin by approximating $I(x,t)$ in (3.33) as $x \to \infty$ with $t = O(1)$, and we find the approximation has three distinct asymptotic regions as $x \to \infty$. We then construct the approximation to $\bar{\eta}(x,t)$, for $t = O(1)$ as $|x| \to \infty$, via (3.32) and (3.34).

5.1   Coordinate Expansion for $I(x,t)$ as $|x| \to \infty$

We examine $I(x,t)$, as given in (3.33) as $x \to \infty$ with $t = O(1)$. We first integrate the integrand in (3.33) around the contour C_Σ, shown in Fig. (5.1), where $C_\Sigma = C_\delta^\Sigma \cup L_1 \cup L_2 \cup C_\varepsilon \cup L_3$ with C_ε having radius $0 < \varepsilon < \frac{\pi}{2}$. The integrand in (3.33) is analytic in the region enclosed by C_Σ, (noting that the branch-cut for $\gamma(k)$, is removable, since cos is an even function). Hence, via the Cauchy Residue Theorem,

$$\int_{C_\Sigma} \frac{1}{k^2} \cos(\gamma(k)t) \exp(-ikx) \, dk = 0, \quad (x,t) \in \mathbb{R}^+ \times \mathbb{R}^+. \tag{5.1}$$

Consider the integral along L_1. Set

$$I_{L_1}(x,t) = \int_{L_1} \frac{1}{k^2} \cos(\gamma(k)t) \exp(-ikx) \, dk, \quad (x,t) \in \mathbb{R}^+ \times \mathbb{R}^+. \tag{5.2}$$

On L_1, k can be written as

$$k = \Sigma + i\tau \quad \text{with} \; -\pi \le \tau \le 0, \tag{5.3}$$

so that

$$\gamma^2(k) = (\Sigma + i\tau) + O(\Sigma \exp(-2\Sigma)) \quad \text{as } \Sigma \to \infty \text{ uniformly for } \tau \in [-\pi, 0],$$

from which

$$\gamma(k) = \Sigma^{\frac{1}{2}} + i\frac{\tau \Sigma^{-\frac{1}{2}}}{2} + O\left(\Sigma^{-\frac{3}{2}}\right) \quad \text{as } \Sigma \to \infty \text{ uniformly for } \tau \in [-\pi, 0]. \tag{5.4}$$

It follows from (5.4) that

$$\left|\cos(\gamma(k)t)\right| \leq \frac{1}{2}\exp\left(-\frac{t\tau}{2}\Sigma^{-\frac{1}{2}}\right) + \frac{1}{2}\exp\left(\frac{t\tau}{2}\Sigma^{-\frac{1}{2}}\right) \leq 2, \qquad (5.5)$$

as $\Sigma \to \infty$ uniformly for $\tau \in [-\pi, 0]$, whilst it is readily verified that, on L_1,

$$\left|\exp(-ikx)\right| = \left|\exp(-i\Sigma x)\exp(\tau x)\right| \leq 1, \quad \text{as } \Sigma \to \infty \text{ uniformly for } \tau \in [-\pi, 0]$$
$$(5.6)$$

and

$$\frac{1}{k^2} = \frac{1}{\Sigma^2} + O\left(\frac{1}{\Sigma^3}\right) \leq \frac{2}{\Sigma^2}, \quad \text{as } \Sigma \to \infty \text{ uniformly for } \tau \in [-\pi, 0]. \qquad (5.7)$$

Thus, via (5.3), (5.5), (5.6) and (5.7), we have

$$\left|\int_{L_1} \frac{1}{k^2}\cos(\gamma(k)t)\exp(-ikx)\,dk\right| \leq \int_{-\pi}^{0} \frac{4}{\Sigma^2}\,d\tau = \frac{4\pi}{\Sigma^2} \to 0, \quad \text{as } \Sigma \to \infty. \quad (5.8)$$

Hence, via (5.2) and (5.8),

$$I_{L_1}(x,t) = \int_{L_1} \frac{1}{k^2}\cos(\gamma(k)t)\exp(-ikx)\,dk \to 0, \text{ as } \Sigma \to \infty \text{ with } (x,t) \in \mathbb{R}^+ \times \mathbb{R}^+.$$
$$(5.9)$$

Similarly, on L_3, we have that

$$I_{L_3}(x,t) = \int_{L_3} \frac{1}{k^2}\cos(\gamma(k)t)\exp(-ikx)\,dk \to 0, \text{ as } \Sigma \to \infty \text{ with } (x,t) \in \mathbb{R}^+ \times \mathbb{R}^+.$$
$$(5.10)$$

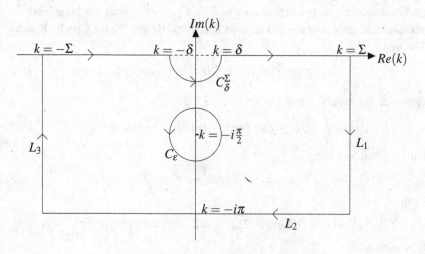

Fig. 5.1: The contour C_Σ in the k-plane.

Now consider the integral along L_2. Set

$$I_{L_2}(x,t) = \int_{L_2} \frac{1}{k^2} \cos(\gamma(k)t) \exp(-ikx)\, dk, \quad (x,t) \in \mathbb{R}^+ \times \mathbb{R}^+. \tag{5.11}$$

On L_2, we write

$$k = \sigma - i\pi \quad \text{with } \sigma \in [-\Sigma, \Sigma], \tag{5.12}$$

so that,

$$\gamma(\sigma - i\pi) = (\sigma - i\pi)^{\frac{1}{2}}(\tanh \sigma)^{\frac{1}{2}}, \quad \sigma \in [-\Sigma, \Sigma]. \tag{5.13}$$

To obtain the correct branches for the square roots in (5.13) we introduce branch cuts in the σ-plane for $(\sigma - i\pi)^{\frac{1}{2}}$ and $(\tanh \sigma)^{\frac{1}{2}}$, as shown in Figs. (5.2) and (5.3), with $\arg\left((\sigma - i\pi)^{\frac{1}{2}}\right) = 0$ on $\sigma = k + i\pi$ when $k > 0$, and $\arg\left((\tanh \sigma)^{\frac{1}{2}}\right) = 0$ with $\sigma > 0$. We observe that

$$(\tanh \sigma)^{\frac{1}{2}} = \begin{cases} 1 + O(\exp(-2\sigma)) & \text{as } \sigma \to \infty, \\ -i + O(\exp(2\sigma)) & \text{as } \sigma \to -\infty, \end{cases}$$

and

$$(\sigma - i\pi)^{\frac{1}{2}} = \begin{cases} \sigma^{\frac{1}{2}}\left(1 - i\dfrac{\pi}{2\sigma}\right) + O\left(\sigma^{-\frac{3}{2}}\right) & \text{as } \sigma \to \infty, \\ (-i)(-\sigma)^{\frac{1}{2}}\left(1 - i\dfrac{\pi}{2\sigma}\right) + O\left((-\sigma)^{-\frac{3}{2}}\right) & \text{as } \sigma \to -\infty, \end{cases}$$

which gives, from (5.13),

$$\gamma(\sigma - i\pi) = \begin{cases} \sigma^{\frac{1}{2}}\left(1 - i\dfrac{\pi}{2\sigma}\right) + O\left(\sigma^{-\frac{3}{2}}\right) & \text{as } \sigma \to \infty, \\ -(-\sigma)^{\frac{1}{2}}\left(1 - i\dfrac{\pi}{2\sigma}\right) + O\left((-\sigma)^{-\frac{3}{2}}\right) & \text{as } \sigma \to -\infty. \end{cases}$$

It now follows from (5.11) and (5.12) that

$$I_{L_2}(x,t) = -\int_{-\Sigma}^{\Sigma} \frac{1}{(\sigma - i\pi)^2} \cos(\gamma(\sigma - i\pi)t) \exp(-i(\sigma - i\pi)x)\, d\sigma. \tag{5.14}$$

From (5.1) we have

$$\int_{C_\delta^\Sigma} \frac{1}{k^2} \cos(\gamma(k)t) \exp(-ikx)\, dk = -I_{L_1}(x,t) - I_{L_2}(x,t) - I_{L_3}(x,t)$$

$$\tag{5.15}$$

$$- \int_{C_\epsilon} \frac{1}{k^2} \cos(\gamma(k)t) \exp(-ikx)\, dk$$

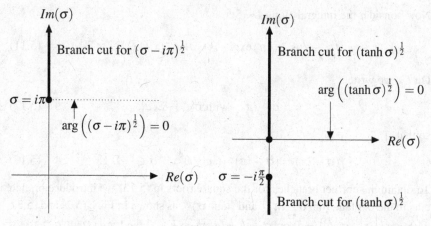

Fig. 5.2: The σ-plane for $(\sigma - i\pi)^{\frac{1}{2}}$.   Fig. 5.3: The σ-plane for $(\tanh \sigma)^{\frac{1}{2}}$.

for any $\Sigma > 0$ and $(x,t) \in \mathbb{R}^+ \times \mathbb{R}^+$. Now let $\Sigma \to \infty$ in (5.15), then via (3.33), (5.9), (5.10) and (5.14), we have

$$I(x,t) = -\int_{C_\varepsilon} \frac{1}{k^2} \cos(\gamma(k)t) \exp(-ikx) \, dk$$

$$+ \int_{-\infty}^{\infty} \frac{1}{(\sigma - i\pi)^2} \cos(\gamma(\sigma - i\pi)t) \exp(-i(\sigma - i\pi)x) \, d\sigma \tag{5.16}$$

for $(x,t) \in \mathbb{R}^+ \times \mathbb{R}^+$. We now estimate the second term on the right hand side of (5.16). We have,

$$\left| \int_{-\infty}^{\infty} \frac{1}{(\sigma - i\pi)^2} \cos(\gamma(\sigma - i\pi)t) \exp(-i(\sigma - i\pi)x) \, d\sigma \right|$$

$$\leq \frac{\exp(-\pi x)}{2} \int_{-\infty}^{\infty} \frac{1}{(\sigma^2 + \pi^2)} \left(|\exp(i\gamma(\sigma - i\pi)t)| + |\exp(-i\gamma(\sigma - i\pi)t)| \right) d\sigma. \tag{5.17}$$

In order to find a bound for (5.17) we establish bounds for each of the exponential terms. First consider $|\exp(i\gamma(\sigma - i\pi)t)|$ for $\sigma \in \mathbb{R}$. It follows from (5.13) that

$$|\mathrm{Im}(\gamma(\sigma - i\pi))| \leq \frac{\pi}{\sqrt{2}\left(\sigma + (\sigma^2 + \pi^2)^{\frac{1}{2}}\right)^{\frac{1}{2}}} \quad \text{for } \sigma \in \mathbb{R}.$$

Thus,

$$|\mathrm{Im}(\gamma(\sigma - i\pi))| \leq \sqrt{\frac{\pi}{2}} \quad \text{for } \sigma \in \mathbb{R}. \tag{5.18}$$

Therefore, we have

$$|\exp(i\gamma(\sigma - i\pi)t))| = \exp(-\text{Im}(\gamma(\sigma - i\pi))t)$$

$$\leq \exp\left(\sqrt{\frac{\pi}{2}}t\right) \tag{5.19}$$

for $\sigma \in \mathbb{R}$, via (5.18). Similarly,

$$|\exp(-i\gamma(\sigma - i\pi)t))| \leq \exp\left(\sqrt{\frac{\pi}{2}}t\right) \tag{5.20}$$

for $\sigma \in \mathbb{R}$. Thus, via (5.17) (5.19) and (5.20) we now have

$$\left|\int_{-\infty}^{\infty} \frac{1}{(\sigma - i\pi)^2} \cos(\gamma(\sigma - i\pi)t) \exp(-i(\sigma - i\pi)x)\,d\sigma\right|$$

$$\tag{5.21}$$

$$\leq \exp(-\pi x)\exp\left(\sqrt{\frac{\pi}{2}}t\right)\int_{-\infty}^{\infty} \frac{1}{\sigma^2 + \pi^2}\,d\sigma, \quad (x,t) \in \mathbb{R}^+ \times \mathbb{R}^+.$$

It is readily established that

$$\int_{-\infty}^{\infty} \frac{1}{\sigma^2 + \pi^2}\,d\sigma = 1,$$

and so, via (5.21), we have

$$\left|\int_{-\infty}^{\infty} \frac{1}{(\sigma - i\pi)^2} \cos(\gamma(\sigma - i\pi)t) \exp(-i(\sigma - i\pi)x)\,d\sigma\right|$$

$$\leq \exp(-\pi x)\exp\left(\sqrt{\frac{\pi}{2}}t\right), \quad (x,t) \in \mathbb{R}^+ \times \mathbb{R}^+.$$

It now follows from (5.16) that

$$I(x,t) = -\int_{C_\varepsilon} \frac{1}{k^2} \cos(\gamma(k)t) \exp(-ikx)\,dk + R(x,t), \quad \text{for } (x,t) \in \mathbb{R}^+ \times \mathbb{R}^+,$$

$$\tag{5.22}$$

with

$$|R(x,t)| \leq \exp\left(-\pi x + \sqrt{\frac{\pi}{2}}t\right), \quad \text{for } (x,t) \in \mathbb{R}^+ \times \mathbb{R}^+. \tag{5.23}$$

Now consider the integral around C_ε in (5.22). Let

$$I_{C_\varepsilon}(x,t) = \int_{C_\varepsilon} \frac{1}{k^2} \cos(\gamma(k)t) \exp(-ikx)\,dk, \quad (x,t) \in \mathbb{R}^+ \times \mathbb{R}^+. \tag{5.24}$$

On C_ε we write

$$k = -i\frac{\pi}{2} + \bar{k}, \quad \text{with } |\bar{k}| = \varepsilon, \tag{5.25}$$

and we will work in the $\bar{k}$-plane, as shown in Fig. (5.4). In a neighbourhood of $\bar{k} = 0$ we have, via Laurent's theorem,

$$\gamma^2\left(\bar{k}\right) = -i\frac{\pi}{2\bar{k}}\left(1 + \frac{2i}{\pi}\bar{k} + \frac{1}{3}\bar{k}^2 + \frac{2i}{3\pi}\bar{k}^3 + O\left(\bar{k}^4\right)\right).$$

We need to correctly choose the square root for $\gamma(\bar{k})$. Recalling the k-plane, see Fig. (3.1), we have defined $\arg(k)$ such that

$$\arg(k) = 0, \text{ when } k \in \mathbb{R}^+,$$

$$\arg(k) = -\frac{\pi}{2}, \text{ when } k = i\tau, \ \tau \in \left(-\frac{\pi}{2}, 0\right),$$

which leads to

$$\arg(\gamma(k)) = 0 \text{ on } k \in \mathbb{R}^+,$$

$$\arg(\gamma(k)) = -\frac{\pi}{2} \text{ on } k = i\tau, \ \tau \in \left(-\frac{\pi}{2}, 0\right). \tag{5.26}$$

Consider $k = i\tau$, for $\tau \in \left(-\frac{\pi}{2}, 0\right)$. Then via (5.26), we have

$$\gamma(k) = -i|\gamma(k)|$$

in the k-plane. Hence, in the $\bar{k}$-plane, we must have $\gamma\left(\bar{k}\right) = -i\left|\gamma\left(\bar{k}\right)\right|$ when $\arg\left(\bar{k}\right) = \frac{\pi}{2}$. Hence, we define $\bar{k}^{\frac{1}{2}}$ to be real and positive with $\bar{k} > 0$, and with

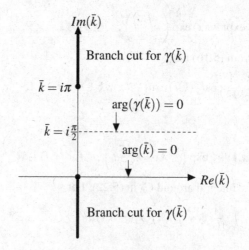

Fig. 5.4: Cut $\bar{k}$-plane.

a branch-cut along the negative imaginary $\bar{k}$-axis. With this definition, we then have,

$$\gamma(\bar{k}) = \sqrt{\frac{\pi}{2}} \exp\left(-i\frac{\pi}{4}\right) \frac{1}{\bar{k}^{\frac{1}{2}}} \left(1 + \frac{2i}{\pi}\bar{k} + \frac{1}{3}\bar{k}^2 + \frac{2i}{3\pi}\bar{k}^3 + O\left(\bar{k}^4\right)\right)^{\frac{1}{2}} \quad \text{as } \bar{k} \to 0. \tag{5.27}$$

As $\bar{k} \to 0$, we may further expand (5.27) to obtain

$$\gamma(\bar{k}) = \sqrt{\frac{\pi}{2}} \exp\left(-i\frac{\pi}{4}\right) \frac{1}{\bar{k}^{\frac{1}{2}}} \left(1 + \frac{i}{\pi}\bar{k} + \frac{1}{2}\left(\frac{1}{3} + \frac{1}{\pi^2}\right)\bar{k}^2 + O\left(\bar{k}^3\right)\right) \quad \text{as } \bar{k} \to 0.$$

Moving into the $\bar{k}$-plane, via (5.24) and (5.25), we obtain

$$I_{C_\varepsilon}(x,t) = -4\exp\left(-\frac{\pi}{2}x\right) \int_{C_\varepsilon} \frac{1}{(\pi + 2i\bar{k})^2} \cos\left(\gamma(\bar{k})t\right) \exp\left(-i\bar{k}x\right) d\bar{k} \tag{5.28}$$

where now C_ε is the circle of radius ε centred at the origin of the $\bar{k}$-plane. We wish to estimate (5.28) as $x \to \infty$ with $t = O(1)$. To facilitate this, we now set

$$\varepsilon = x^{-\frac{2}{3}}$$

in (5.28), after which we have,

$$I_{C_\varepsilon}(x,t) = -4\exp\left(-\frac{\pi}{2}x\right) \left(\int_{C_{x^{-\frac{2}{3}}}} \frac{1}{(\pi + 2i\bar{k})^2} \cos\left((\gamma(\bar{k})t)\exp\left(-i\bar{k}x\right)\right) d\bar{k}\right) \tag{5.29}$$

for $x \in \left(\left(\frac{2}{\pi}\right)^{\frac{3}{2}}, \infty\right)$ and $t \geq 0$, recalling that $0 < \varepsilon < \frac{1}{2}\pi$. We now write (5.29) as,

$$I_{C_\varepsilon}(x,t) = -2\exp\left(-\frac{\pi}{2}x\right) \left(\int_{C_{x^{-\frac{2}{3}}}} \frac{1}{(\pi + 2i\bar{k})^2} \exp\left(i\left(\gamma(\bar{k})t - \bar{k}x\right)\right) d\bar{k}\right.$$

$$\left. + \int_{C_{x^{-\frac{2}{3}}}} \frac{1}{(\pi + 2i\bar{k})^2} \exp\left(-i\left(\gamma(\bar{k})t + \bar{k}x\right)\right) d\bar{k}\right), \tag{5.30}$$

noting that the branch cuts for $\gamma(\bar{k})$ on the imaginary $\bar{k}$ axis must now be restored in each separate integrand. It is now convenient to make the substitution

$$s = x^{\frac{2}{3}}\bar{k}$$

in the integrals in (5.30), which becomes

$$I_{C_\varepsilon}(x,t) = -\frac{2}{x^{\frac{2}{3}}} \exp\left(-\frac{\pi}{2}x\right) \left(\int_{C_1} \frac{1}{\left(\pi + 2ix^{-\frac{2}{3}}s\right)^2} \exp\left(i\left(\gamma\left(x^{-\frac{2}{3}}s\right)t - x^{\frac{1}{3}}s\right)\right) ds\right.$$

$$\left. + \int_{C_1} \frac{1}{\left(\pi + 2ix^{-\frac{2}{3}}s\right)^2} \exp\left(-i\left(\gamma\left(x^{-\frac{2}{3}}s\right)t + x^{\frac{1}{3}}s\right)\right) ds\right) \tag{5.31}$$

for $x \in \left(\left(\frac{2}{\pi} \right)^{\frac{3}{2}}, \infty \right)$ and $t \geq 0$. On the contour C_1, we have $|s| = 1$, and we may expand as follows,

$$\left(\pi + 2ix^{-\frac{2}{3}}s \right)^{-2} = \frac{1}{\pi^2} \left(1 - i\frac{4}{\pi}x^{-\frac{2}{3}}s + O\left(x^{-\frac{4}{3}}s^2 \right) \right), \tag{5.32}$$

$$\exp \left(i \left(\gamma \left(x^{-\frac{2}{3}}s \right) t - x^{\frac{1}{3}}s \right) \right)$$

$$= \exp \left(i \left(t\sqrt{\frac{\pi}{2}} e^{-i\frac{\pi}{4}} \left(x^{\frac{1}{3}}s^{-\frac{1}{2}} + \frac{i}{\pi}x^{-\frac{1}{3}}s^{\frac{1}{2}} + O\left(x^{-1}s^{\frac{3}{2}} \right) \right) - x^{\frac{1}{3}}s \right) \right)$$

$$= \exp \left(i \left(\sqrt{\frac{\pi}{2}} e^{-i\frac{\pi}{4}} t \left(x^{\frac{1}{3}}s^{-\frac{1}{2}} + \frac{i}{\pi}x^{-\frac{1}{3}}s^{\frac{1}{2}} \right) - x^{\frac{1}{3}}s \right) \right) \left(1 + O\left(tx^{-1}s^{\frac{3}{2}} \right) \right), \tag{5.33}$$

$$\exp \left(-i \left(\gamma \left(x^{-\frac{2}{3}}s \right) t + x^{\frac{1}{3}}s \right) \right)$$

$$= \exp \left(-i \left(t\sqrt{\frac{\pi}{2}} e^{-i\frac{\pi}{4}} \left(x^{\frac{1}{3}}s^{-\frac{1}{2}} + \frac{i}{\pi}x^{-\frac{1}{3}}s^{\frac{1}{2}} + O\left(x^{-1}s^{\frac{3}{2}} \right) \right) + x^{\frac{1}{3}}s \right) \right)$$

$$= \exp \left(-i \left(\sqrt{\frac{\pi}{2}} e^{-i\frac{\pi}{4}} t \left(x^{\frac{1}{3}}s^{-\frac{1}{2}} + \frac{i}{\pi}x^{-\frac{1}{3}}s^{\frac{1}{2}} \right) + x^{\frac{1}{3}}s \right) \right) \left(1 + O\left(tx^{-1}s^{\frac{3}{2}} \right) \right) \tag{5.34}$$

as $x \to \infty$ with $t \geq 0$ and $|s| = 1$. Writing (5.32), (5.33) and (5.34) in (5.31) leads to

$$I_{C_\varepsilon}(x,t) = -\frac{2}{\pi^2 x^{\frac{2}{3}}} \exp \left(-\frac{\pi}{2}x \right) \times \left(\int_{C_1} \exp \left(i \left(\sqrt{\frac{\pi}{2}} e^{-i\frac{\pi}{4}} t \left(x^{\frac{1}{3}}s^{-\frac{1}{2}} + \frac{i}{\pi}x^{-\frac{1}{3}}s^{\frac{1}{2}} \right. \right. \right. \right.$$

$$\left. \left. \left. - x^{\frac{1}{3}}s \right) \right) \times \left(1 + O\left(\frac{s}{x^{\frac{2}{3}}} + \frac{ts^{\frac{3}{2}}}{x} \right) \right) ds + \int_{C_1} \exp \left(-i \left(\sqrt{\frac{\pi}{2}} e^{-i\frac{\pi}{4}} t \left(x^{\frac{1}{3}}s^{-\frac{1}{2}} \right. \right. \right. \right.$$

$$\left. \left. \left. \left. + \frac{i}{\pi}x^{-\frac{1}{3}}s^{\frac{1}{2}} \right) + x^{\frac{1}{3}}s \right) \right) \times \left(1 + O\left(\frac{s}{x^{\frac{2}{3}}} + \frac{ts^{\frac{3}{2}}}{x} \right) \right) ds \right),$$

as $x \to \infty$ with $t = O(1)$. Hence,

$$I_{C_\varepsilon}(x,t) = -\frac{4}{\pi^2 x^{\frac{2}{3}}} \exp\left(-\frac{\pi}{2}x\right) \int_{C_1} \cos\left(\sqrt{\frac{\pi}{2}} e^{-i\frac{\pi}{4}} t \left(x^{\frac{1}{3}} s^{-\frac{1}{2}} + \frac{i}{\pi} x^{-\frac{1}{3}} s^{\frac{1}{2}}\right)\right)$$

$$\times \exp\left(-ix^{\frac{1}{3}} s\right) ds + O\left(\frac{1}{x^{\frac{2}{3}}} \exp\left(-\frac{\pi}{2}x\right) \int_{C_1} \left(\frac{s}{x^{\frac{2}{3}}} + \frac{ts^{\frac{3}{2}}}{x}\right)\right)$$

$$\times \cos\left(\sqrt{\frac{\pi}{2}} e^{-i\frac{\pi}{4}} t \left(x^{\frac{1}{3}} s^{-\frac{1}{2}} + \frac{i}{\pi} x^{-\frac{1}{3}} s^{\frac{1}{2}}\right)\right) \exp\left(-ix^{\frac{1}{3}} s\right) ds$$

as $x \to \infty$ with $t = O(1)$. Thus we have

$$I_{C_\varepsilon}(x,t) = -\frac{4}{\pi^2 x^{\frac{2}{3}}} \exp\left(-\frac{\pi}{2}x\right) \tilde{I}(x,t) \left(1 + O\left(\frac{1}{x^{\frac{2}{3}}} + \frac{t}{x}\right)\right) \tag{5.35}$$

as $x \to \infty$ with $t = O(1)$. Here

$$\tilde{I}(x,t) = \int_{C_1} \cos\left(\sqrt{\frac{\pi}{2}} e^{-i\frac{\pi}{4}} t \left(x^{\frac{1}{3}} s^{-\frac{1}{2}} + \frac{i}{\pi} x^{-\frac{1}{3}} s^{\frac{1}{2}}\right)\right) \exp\left(-ix^{\frac{1}{3}} s\right) ds \tag{5.36}$$

for $x > 0$ and $t \geq 0$. We now observe that the integrand in (5.36) is analytic for $s \in \mathbb{C} \backslash \{0\}$ (with an essential singularity at $s = 0$). Thus, via Cauchy's Theorem, we can deform the contour in (5.36) from C_1 onto C_R, with

$$R = \frac{1}{2} \pi^{\frac{1}{3}} t^{\frac{2}{3}}.$$

We then have,

$$\tilde{I}(x,R) = \int_{C_R} \cos\left(2e^{-i\frac{\pi}{4}} R^{\frac{3}{2}} \left(x^{\frac{1}{3}} s^{-\frac{1}{2}} + \frac{i}{\pi} x^{-\frac{1}{3}} s^{\frac{1}{2}}\right)\right) \exp\left(-ix^{\frac{1}{3}} s\right) ds$$

for $x > 0$ and $R \geq 0$, which we may write as

$$\tilde{I}(x,R) = \frac{1}{2} \int_{C_R} \exp\left(ix^{\frac{1}{3}} \bar{f}_1(s)\right) ds + \frac{1}{2} \int_{C_R} \exp\left(-ix^{\frac{1}{3}} \bar{f}_2(s)\right) ds \tag{5.37}$$

for $x > 0$ and $R \geq 0$, with

$$\bar{f}_1(s) = 2\exp\left(-i\frac{\pi}{4}\right) R^{\frac{3}{2}} \left(s^{-\frac{1}{2}} + \frac{i}{\pi} x^{-\frac{2}{3}} s^{\frac{1}{2}}\right) - s$$

$$\bar{f}_2(s) = 2\exp\left(-i\frac{\pi}{4}\right) R^{\frac{3}{2}} \left(s^{-\frac{1}{2}} + \frac{i}{\pi} x^{-\frac{2}{3}} s^{\frac{1}{2}}\right) + s,$$

recalling that the branch-cut in each of the integrands in (5.37), along the negative imaginary s-axis, must be restored (with $s^{\frac{1}{2}}$ real and positive on the positive real s-axis). We now make the substitution $s = Rw$ in (5.37) to obtain

$$\tilde{I}(x,\tau) = \frac{\pi^{\frac{1}{3}}}{4}\left(x^{\frac{1}{4}}\tau\right)^{\frac{2}{3}}\int_{C_1} g_1(w,\tau)\exp\left(i\left(x\tau^{\frac{4}{3}}\right)^{\frac{1}{2}}\hat{f}_1(w)\right)dw$$

$$+\frac{\pi^{\frac{1}{3}}}{4}\left(x^{\frac{1}{4}}\tau\right)^{\frac{2}{3}}\int_{C_1} g_2(w,\tau)\exp\left(-i\left(x\tau^{\frac{4}{3}}\right)^{\frac{1}{2}}\hat{f}_2(w)\right)dw$$

for $x > 0$, $\tau \geq 0$, where now,

$$\hat{f}_1(w) = \frac{\pi^{\frac{1}{3}}}{2}\left(2\exp\left(-i\frac{\pi}{4}\right)w^{-\frac{1}{2}} - w\right)$$

$$\hat{f}_2(w) = \frac{\pi^{\frac{1}{3}}}{2}\left(2\exp\left(-i\frac{\pi}{4}\right)w^{-\frac{1}{2}} + w\right)$$

and

$$g_1(w,\tau) = \exp\left(-\frac{1}{2\pi^{\frac{1}{3}}}\exp\left(-i\frac{\pi}{4}\right)\tau^{\frac{4}{3}}w^{\frac{1}{2}}\right)$$

$$g_2(w,\tau) = \exp\left(\frac{1}{2\pi^{\frac{1}{3}}}\exp\left(-i\frac{\pi}{4}\right)\tau^{\frac{4}{3}}w^{\frac{1}{2}}\right)$$

with

$$\tau = x^{-\frac{1}{4}}t.$$

We can now write,

$$\tilde{I}(x,t) = \frac{\pi^{\frac{1}{3}}}{4}t^{\frac{2}{3}}\left(F_1\left(\left(\frac{t}{x^{\frac{1}{4}}}\right),\left(x^{\frac{1}{2}}t\right)^{\frac{2}{3}}\right) + F_2\left(\left(\frac{t}{x^{\frac{1}{4}}}\right),\left(x^{\frac{1}{2}}t\right)^{\frac{2}{3}}\right)\right)$$

for $x > 0$, $t \geq 0$, with,

$$F_1(T,X) = \int_{C_1} g_1(w,T)\exp\left(iX\hat{f}_1(w)\right)dw \tag{5.38}$$

$$F_2(T,X) = \int_{C_1} g_2(w,T)\exp\left(-iX\hat{f}_2(w)\right)dw \tag{5.39}$$

for $X, T \geq 0$. On C_1, we write

$$w = \exp(i\theta) \quad \text{with} -\frac{\pi}{2} \leq \theta \leq \frac{3\pi}{2},$$

which conforms with the branch-cut along $\theta = -\frac{\pi}{2}$. We may then write (5.38) and (5.39) as,

$$F_1(T,X) = i\int_{-\frac{\pi}{2}}^{\frac{3\pi}{2}} g_1(\theta,T)\exp\left(iX\hat{f}_1(\theta)\right)\exp(i\theta)d\theta \tag{5.40}$$

$$F_2(T,X) = i \int_{-\frac{\pi}{2}}^{\frac{3\pi}{2}} g_2(\theta,T) \exp\left(-iX\hat{f}_2(\theta)\right) \exp(i\theta)\, d\theta \tag{5.41}$$

for $X, T \geq 0$, where

$$g_1(\theta,T) = \exp\left(-\frac{1}{2\pi^{\frac{1}{3}}} \exp\left(-i\left(\frac{\pi}{4} - \frac{\theta}{2}\right)\right) T^{\frac{4}{3}}\right) \tag{5.42}$$

$$g_2(\theta,T) = \exp\left(\frac{1}{2\pi^{\frac{1}{3}}} \exp\left(-i\left(\frac{\pi}{4} - \frac{\theta}{2}\right)\right) T^{\frac{4}{3}}\right) \tag{5.43}$$

and

$$\hat{f}_1(\theta) = \frac{\pi^{\frac{1}{3}}}{2}\left(2\exp\left(-i\left(\frac{\pi}{4} + \frac{\theta}{2}\right)\right) - \exp(i\theta)\right) \tag{5.44}$$

$$\hat{f}_2(\theta) = \frac{\pi^{\frac{1}{3}}}{2}\left(2\exp\left(-i\left(\frac{\pi}{4} + \frac{\theta}{2}\right)\right) + \exp(i\theta)\right) \tag{5.45}$$

Finally, setting

$$\theta = \bar{\theta} - \frac{\pi}{2}$$

in (5.40)–(5.45), we obtain,

$$\tilde{I}(x,t) = \frac{\pi^{\frac{1}{3}}}{4} t^{\frac{2}{3}}\left(F_1\left(\left(\frac{t}{x^{\frac{1}{4}}}\right), \left(x^{\frac{1}{2}}t\right)^{\frac{2}{3}}\right) + F_2\left(\left(\frac{t}{x^{\frac{1}{4}}}\right), \left(x^{\frac{1}{2}}t\right)^{\frac{2}{3}}\right)\right) \tag{5.46}$$

for $x > 0$, $t \geq 0$ where

$$F_1(T,X) = \int_0^{2\pi} g_1(\bar{\theta},T) \exp\left(iX\hat{f}_1(\bar{\theta})\right) \exp(i\bar{\theta})\, d\bar{\theta} \tag{5.47}$$

$$F_2(T,X) = \int_0^{2\pi} g_2(\bar{\theta},T) \exp\left(-iX\hat{f}_2(\bar{\theta})\right) \exp(i\bar{\theta})\, d\bar{\theta} \tag{5.48}$$

for $X, T \geq 0$, with

$$g_1(\bar{\theta},T) = \exp\left(\frac{i}{2\pi^{\frac{1}{3}}} \exp\left(i\frac{\bar{\theta}}{2}\right) T^{\frac{4}{3}}\right) \tag{5.49}$$

$$g_2(\bar{\theta},T) = \exp\left(-\frac{i}{2\pi^{\frac{1}{3}}} \exp\left(i\frac{\bar{\theta}}{2}\right) T^{\frac{4}{3}}\right), \tag{5.50}$$

and

$$\hat{f}_1(\bar{\theta}) = \frac{\pi^{\frac{1}{3}}}{2}\left(2\exp\left(-i\frac{\bar{\theta}}{2}\right) + i\exp(i\bar{\theta})\right) \tag{5.51}$$

$$\hat{f}_2(\bar{\theta}) = \frac{\pi^{\frac{1}{3}}}{2}\left(2\exp\left(-i\frac{\bar{\theta}}{2}\right) - i\exp(i\bar{\theta})\right). \tag{5.52}$$

It is now instructive at this stage to verify that $\tilde{I}(x,t)$, $x > 0$, $t \geq 0$, is real valued. It is sufficient to show that $F_1(T,X)$ and $F_2(T,X)$ are real valued for $X,T \geq 0$. To this end we write, via (5.47), (5.49) and (5.51),

$$F_1(T,X) = \int_0^{2\pi} \Re(T,X,\bar{\theta})\,d\bar{\theta} + i\int_0^{2\pi} \Im(T,X,\bar{\theta})\,d\bar{\theta} \tag{5.53}$$

for $X,T \geq 0$, where

$$\Re(T,X,\bar{\theta}) = \exp\left(-X\frac{\pi^{\frac{1}{3}}}{2}\cos\bar{\theta} - \left(\frac{1}{2\pi^{\frac{1}{3}}}T^{\frac{4}{3}} - X\pi^{\frac{1}{3}}\right)\sin\frac{\bar{\theta}}{2}\right)$$

$$\times \cos\left(\left(\frac{1}{2\pi^{\frac{1}{3}}}T^{\frac{4}{3}} + X\pi^{\frac{1}{3}}\right)\cos\frac{\bar{\theta}}{2} + \bar{\theta}\right),$$

$$\Im(T,X,\bar{\theta}) = \exp\left(-X\frac{\pi^{\frac{1}{3}}}{2}\cos\bar{\theta} - \left(\frac{1}{2\pi^{\frac{1}{3}}}T^{\frac{4}{3}} - X\pi^{\frac{1}{3}}\right)\sin\frac{\bar{\theta}}{2}\right)$$

$$\times \sin\left(\left(\frac{1}{2\pi^{\frac{1}{3}}}T^{\frac{4}{3}} + X\pi^{\frac{1}{3}}\right)\cos\frac{\bar{\theta}}{2} + \bar{\theta}\right).$$

It can be verified that

$$\Im(T,X,2\pi - \bar{\theta}) = -\Im(T,X,\bar{\theta})$$

for $X,T \geq 0$, and so we write the second integral on the right-hand side of (5.53) as

$$\int_0^{2\pi} \Im(T,X,\bar{\theta})\,d\bar{\theta} = \int_0^{\pi} \Im(T,X,\bar{\theta})\,d\bar{\theta} + \int_{\pi}^{2\pi} \Im(T,X,\bar{\theta})\,d\bar{\theta} \tag{5.54}$$

for $X,T \geq 0$. Setting $\bar{\theta} = 2\pi - \bar{\bar{\theta}}$ in the second integral on the right-hand side of (5.54) we have

$$\int_0^{2\pi} \Im(T,X,\bar{\theta})\,d\bar{\theta} = \int_0^{\pi} \Im(T,X,\bar{\theta})\,d\bar{\theta} - \int_{\pi}^0 \Im\left(T,X,2\pi - \bar{\bar{\theta}}\right)\,d\bar{\bar{\theta}}$$

$$= \int_0^{\pi} \Im(T,X,\bar{\theta})\,d\bar{\theta} - \int_0^{\pi} \Im\left(T,X,\bar{\bar{\theta}}\right)\,d\bar{\bar{\theta}}$$

$$= 0.$$

Thus, via (5.53) and (5.54), we have that $F_1(T,X)$ is real valued for $X,T \geq 0$. Similarly, $F_2(T,X)$ is real valued for $X,T \geq 0$, hence, via (5.46), $\tilde{I}(x,t)$ is real valued for $x > 0$ and $t \geq 0$.

An examination of (5.46) reveals that $\tilde{I}(x,t)$ has three distinct asymptotic forms as $x \to \infty$ with $t \geq 0$, namely

Region I: $t = O\left(x^{-\frac{1}{2}}\right)$ as $x \to \infty$.

Region II: $t = O(1)$ as $x \to \infty$.

Region III: $t = O\left(x^{\frac{1}{4}}\right)$ as $x \to \infty$.

We consider the approximation of $\tilde{I}(x,t)$ in each region in turn.

5.1.1 Coordinate Expansion for $\tilde{I}(x,t)$ with $t = O\left(x^{-\frac{1}{2}}\right)$ as $|x| \to \infty$

In Region I we introduce the scaled coordinate $\tilde{t} = x^{\frac{1}{2}}t$, so that $\tilde{t} = O(1)$ as $x \to \infty$ in Region I. We then have, via (5.46),

$$\tilde{I}(x,\tilde{t}) = \frac{\pi^{\frac{1}{3}}}{4} \frac{\tilde{t}^{\frac{2}{3}}}{x^{\frac{1}{3}}} \left(F_1\left(\frac{\tilde{t}}{x^{\frac{3}{4}}}, \tilde{t}^{\frac{2}{3}}\right) + F_2\left(\frac{\tilde{t}}{x^{\frac{3}{4}}}, \tilde{t}^{\frac{2}{3}}\right) \right) \tag{5.55}$$

for $x > 0$, $\tilde{t} \geq 0$. We now estimate (5.55) with $\tilde{t} = O(1)$ as $x \to \infty$. This requires estimates of $F_1(T,X)$ and $F_2(T,X)$ with $X = O(1)$ as $T \to 0$. We obtain directly from (5.47)–(5.52),

$$F_1(T,X) = H_1(X)\left(1 + O\left(T^{\frac{4}{3}}\right)\right) \tag{5.56}$$

$$F_2(T,X) = H_2(X)\left(1 + O\left(T^{\frac{4}{3}}\right)\right) \tag{5.57}$$

with $X = O(1)$ as $T \to 0$, where

$$H_1(X) = \int_0^{2\pi} \exp\left(iX\hat{f}_1(\bar{\theta})\right)\exp(i\bar{\theta})\,d\bar{\theta} \tag{5.58}$$

$$H_2(X) = \int_0^{2\pi} \exp\left(-iX\hat{f}_2(\bar{\theta})\right)\exp(i\bar{\theta})\,d\bar{\theta} \tag{5.59}$$

for $X \geq 0$. Graphs of the numerical calculations of $H_1(X)$ and $H_2(X)$, for $X \geq 0$, are given in Figs. (5.5) and (5.6). It is useful, at this stage, to obtain the approximate forms for both $H_1(X)$ and $H_2(X)$ for $0 \leq X \ll 1$ and $X \gg 1$ respectively. For $0 \leq X \ll 1$, we have from (5.58) and (5.59) that

$$H_1(X) = \int_0^{2\pi} \left(1 + iX\hat{f}_1(\bar{\theta}) - \frac{1}{2}X^2\hat{f}_1^2(\bar{\theta}) - i\frac{1}{6}X\hat{f}_1^3(\bar{\theta}) + \frac{1}{24}X^4\hat{f}_1^4(\bar{\theta})\right.$$

$$\left. + i\frac{1}{120}X^5\hat{f}_1^5(\bar{\theta}) + O(X^6) \right)\exp(i\bar{\theta})\,d\bar{\theta}$$

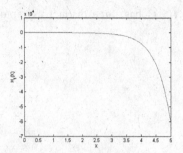

Fig. 5.5: Graph of $H_1(X)$, $X \geq 0$.   Fig. 5.6: Graph of $H_2(X)$, $X \geq 0$.

and

$$H_2(X) = \int_0^{2\pi} \left(1 - iX\hat{f}_2(\bar{\theta}) - \frac{1}{2}X^2\hat{f}_2^2(\bar{\theta}) + i\frac{1}{6}X\hat{f}_2^3(\bar{\theta}) + \frac{1}{24}X^4\hat{f}_2^4(\bar{\theta}) \right.$$

$$\left. -i\frac{1}{120}X^5\hat{f}_2^5(\bar{\theta}) + O(X^6) \right) \exp(i\bar{\theta})\, d\bar{\theta}$$

for $0 \leq X \ll 1$, from which, via (5.51) and (5.52), we find

$$H_1(X) + H_2(X) = -\int_0^{2\pi} \pi^{\frac{2}{3}}X^2 + O\left(X^5\right)\, d\bar{\theta}$$

$$= -2\pi^{\frac{5}{3}}X^2 + O\left(X^5\right)$$

(5.60)

for $0 \leq X \ll 1$. The graph of (5.60) with the numerical calculation for $H_1(X) + H_2(X)$ is shown in Fig. (5.7).

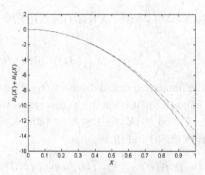

Fig. 5.7: The graph of $H_1(X) + H_2(X)$, with asymptotic approximation $(--)$ from (5.60), for $0 \leq X \ll 1$.

For $X \gg 1$, $H_1(X)$ and $H_2(X)$ are both steepest descent type integrals. Thus we write, from (5.51), (5.52), (5.58) and (5.59),

$$H_1(X) = \int_0^{2\pi} \exp\left(\frac{\pi^{\frac{1}{3}}}{2}X\left(Re_1(\bar{\theta}) - iIm_1(\bar{\theta})\right)\right) \exp(i\bar{\theta})\,d\bar{\theta} \tag{5.61}$$

and

$$H_2(X) = \int_0^{2\pi} \exp\left(\frac{\pi^{\frac{1}{3}}}{2}X\left(Re_2(\bar{\theta}) - iIm_2(\bar{\theta})\right)\right) \exp(i\bar{\theta})\,d\bar{\theta} \tag{5.62}$$

for $X \geq 0$, with

$$Re_1(\bar{\theta}) = 2\sin\left(\frac{\bar{\theta}}{2}\right) - \cos(\bar{\theta}),$$

$$Re_2(\bar{\theta}) = -2\sin\left(\frac{\bar{\theta}}{2}\right) - \cos(\bar{\theta}),$$

and

$$Im_1(\bar{\theta}) = 2\cos\left(\frac{\bar{\theta}}{2}\right) - \sin(\bar{\theta}),$$

$$Im_2(\bar{\theta}) = 2\cos\left(\frac{\bar{\theta}}{2}\right) + \sin(\bar{\theta}),$$

Graphs of $Re_2(\bar{\theta})$ and $Im_2(\bar{\theta})$ are shown in Figs. (5.8) and (5.9) respectively, and it is readily verified that $\max_{\bar{\theta} \in [0,2\pi]}\left(Re_2(\bar{\theta})\right) = -1$. Thus we have from (5.62),

$$|H_2(X)| \leq 2\pi \exp\left(-\frac{\pi^{\frac{1}{3}}}{2}X\right) \tag{5.63}$$

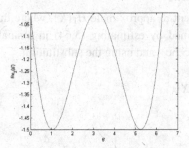

Fig. 5.8: Graph of $Re_2(\bar{\theta})$.

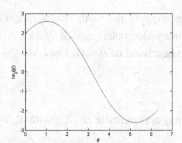

Fig. 5.9: Graph of $Im_2(\bar{\theta})$.

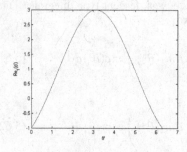
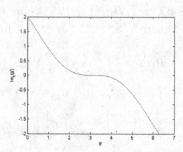

Fig. 5.10: Graph of $Re_1(\bar{\theta})$.   Fig. 5.11: Graph of $Im_1(\bar{\theta})$.

for $X \geq 0$. Graphs of $Re_1(\bar{\theta})$ and $Im_1(\bar{\theta})$ are shown in Figs. (5.10) and (5.11) respectively, and it is readily verified that $\max\limits_{\bar{\theta}\in[0,2\pi]}\left(Re_1(\bar{\theta})\right) = 3$ at $\bar{\theta} = \pi$, and $Im_1(\bar{\theta})$ is stationary at $\bar{\theta} = \pi$. Thus, we anticipate that $H_1(X)$ will be dominated in a small neighbourhood of $\bar{\theta} = \pi$ for $X \gg 1$. It is therefore convenient to write

$$\bar{\theta} = \pi + \tilde{\theta}$$

in (5.61), to obtain

$$H_1(X) = -\int_{-\pi}^{\pi} \exp\left(\frac{\pi^{\frac{1}{3}}}{2}X\left(Re_1(\tilde{\theta}+\pi) - iIm_1(\tilde{\theta}+\pi)\right)\right)\exp(i\tilde{\theta})\,d\tilde{\theta} \quad (5.64)$$

for $X \geq 0$, so that $H_1(X)$ is dominated in a small neighbourhood of $\tilde{\theta} = 0$ for $X \gg 1$. For $\tilde{\theta} \ll 1$ we have

$$Re_1(\tilde{\theta}+\pi) = 3 - \frac{3}{4}\tilde{\theta}^2 + \frac{3}{64}\tilde{\theta}^4 + O\left(\tilde{\theta}^6\right) \quad (5.65)$$

$$Im_1(\tilde{\theta}+\pi) = -\frac{1}{8}\tilde{\theta}^3 + \frac{1}{128}\tilde{\theta}^5 + O\left(\tilde{\theta}^7\right). \quad (5.66)$$

We now apply the method of steepest descent to approximate $H_1(X)$, where the leading order behaviour of $H_1(X)$ is obtained by estimating (5.64) in a small neighbourhood of $\tilde{\theta} = 0$. Thus, via (5.64)–(5.66), and using the substitution

$$u^2 = \frac{3\pi^{\frac{1}{3}}}{8}X\tilde{\theta}^2$$

we may approximate $H_1(X)$, with $X \gg 1$, as

$$H_1(X) = -\frac{2\sqrt{2}}{\sqrt{3}\pi^{\frac{1}{6}}}X^{-\frac{1}{2}}\exp\left(\frac{3\pi^{\frac{1}{3}}}{2}X\right)\int_{-\infty}^{\infty}\exp(-u^2)\left(1+O\left(\frac{u^4}{X}\right)\right)du.$$

Thus we obtain the approximation,

$$H_1(X) = -\frac{2\sqrt{2}\pi^{\frac{1}{3}}}{\sqrt{3}}X^{-\frac{1}{2}}\exp\left(\frac{3\pi^{\frac{1}{3}}}{2}X\right)\left(1+O\left(\frac{1}{X}\right)\right) \qquad (5.67)$$

for $X \gg 1$. It then follows, via (5.63) and (5.67), that

$$H_1(X)+H_2(X) = -\frac{2\sqrt{2}\pi^{\frac{1}{3}}}{\sqrt{3}}X^{-\frac{1}{2}}\exp\left(\frac{3\pi^{\frac{1}{3}}}{2}X\right)\left(1+O\left(\frac{1}{X}\right)\right) \qquad (5.68)$$

for $X \gg 1$. The graph of (5.68) with the numerical calculation for $H_1(X)+H_2(X)$ is shown in Fig. (5.12), and the graph of the ratio between the numerical calculation and the asymptotic approximation (5.68) shown in Fig. (5.13). On returning to (5.55), with (5.56) and (5.57), we have,

$$\tilde{I}(x,\tilde{t}) = \frac{\pi^{\frac{1}{3}}}{4}\frac{\tilde{t}^{\frac{2}{3}}}{x^{\frac{1}{3}}}\left(H_1\left(\tilde{t}^{\frac{2}{3}}\right)+H_2\left(\tilde{t}^{\frac{2}{3}}\right)\right)\left(1+O\left(\frac{\tilde{t}^{\frac{4}{3}}}{x}\right)\right) \qquad (5.69)$$

with $\tilde{t} = O(1)$ as $x \to \infty$. In addition, via (5.60) and (5.68),

$$\tilde{I}(x,\tilde{t}) = -\frac{1}{2}\pi^2\frac{\tilde{t}^2}{x^{\frac{1}{3}}}\left(1+O\left(\tilde{t}^2\right)\right)\left(1+O\left(\frac{\tilde{t}^{\frac{4}{3}}}{x}\right)\right)$$

for $0 \le \tilde{t} \ll 1$ as $x \to \infty$, and

$$\tilde{I}(x,\tilde{t}) = -\frac{\pi^{\frac{2}{3}}}{\sqrt{6}}\frac{\tilde{t}^{\frac{1}{3}}}{x^{\frac{1}{3}}}\exp\left(\frac{3\pi^{\frac{1}{3}}}{2}\tilde{t}^{\frac{2}{3}}\right)\left(1+O\left(\frac{1}{\tilde{t}^{\frac{2}{3}}}\right)\right)\left(1+O\left(\frac{\tilde{t}^{\frac{4}{3}}}{x}\right)\right)$$

for $\tilde{t} \gg 1$ as $x \to \infty$.

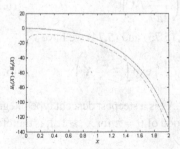 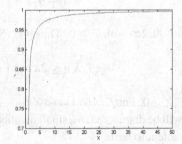

Fig. 5.12: The graph of $H_1(X) + H_2(X)$, with asymptotic approximation ($--$) from (5.68), for $X \gg 1$.

Fig. 5.13: Graph of the ratio between the numerical approximation for $H_1(X) + H_2(X)$ and the asymptotic approximation from (5.68).

5.1.2   *Coordinate Expansion for $\tilde{I}(x,t)$ with $t \geq O(1)$ as $|x| \to \infty$*

In Region II we have $t = O(1)$ as $x \to \infty$. This requires estimates of $F_1(T,X)$ and $F_2(T,X)$ with $0 < T \ll 1$ and $X \gg 1$. Following (5.56), (5.57), (5.63) and (5.67), we have,

$$F_1(T,X) + F_2(T,X) = -\frac{2\sqrt{2}\pi^{\frac{1}{3}}}{\sqrt{3}} X^{-\frac{1}{2}} \exp\left(\frac{3\pi^{\frac{1}{3}}}{2}X\right)\left(1 + O\left(\frac{1}{X}\right)\right)\left(1 + O\left(T^{\frac{4}{3}}\right)\right)$$

(5.70)

with $0 < T \ll 1$ and $X \gg 1$. It then follows from (5.46) and (5.70), that

$$\tilde{I}(x,t) = -\frac{\pi^{\frac{2}{3}} t^{\frac{1}{3}}}{\sqrt{6}\, x^{\frac{1}{6}}} \exp\left(\frac{3\pi^{\frac{1}{3}}}{2}x^{\frac{1}{3}}t^{\frac{2}{3}}\right)\left(1 + O\left(\frac{1}{x^{\frac{1}{3}}t^{\frac{2}{3}}}\right)\right)\left(1 + O\left(\frac{t^{\frac{4}{3}}}{x^{\frac{1}{3}}}\right)\right) \quad (5.71)$$

with $t = O(1)$ as $x \to \infty$.

In Region III we introduce the scaled coordinate $\hat{t} = x^{-\frac{1}{4}}t$, so that $\hat{t} = O(1)$ as $x \to \infty$ in this region. We then have, via (5.46),

$$\tilde{I}(x,\hat{t}) = \frac{\pi^{\frac{1}{3}}}{4} x^{\frac{1}{6}}\hat{t}^{\frac{2}{3}}\left(F_1\left(\hat{t}, \left(x^{\frac{3}{4}}\hat{t}\right)^{\frac{2}{3}}\right) + F_2\left(\hat{t}, \left(x^{\frac{3}{4}}\hat{t}\right)^{\frac{2}{3}}\right)\right) \quad (5.72)$$

for $x > 0$, $\hat{t} \geq 0$. We now estimate (5.72) with $\hat{t} = O(1)$ as $x \to \infty$. This requires estimates of $F_1(T,X)$ and $F_2(T,X)$ with $T = O(1)$ and $X \gg 1$. It is readily verified from (5.50) and (5.52), that

$$\left|\exp\left(iXf_2(\bar{\theta})\right)\right| \leq \exp\left(-\frac{\pi^{\frac{1}{3}}}{2}X\right) \quad (5.73)$$

for $\bar{\theta} \in [0, 2\pi]$ with $X \geq 0$, and

$$\left|g_2(\bar{\theta}, T)\right| \leq \exp\left(\frac{1}{2\pi^{\frac{1}{3}}}T^{\frac{4}{3}}\right) \quad (5.74)$$

for $\bar{\theta} \in [0, 2\pi]$ with $T \geq 0$. Thus, via (5.48), (5.73) and (5.74) we have

$$\left|F_2(T,X)\right| \leq 2\pi \exp\left(\frac{1}{2\pi^{\frac{1}{3}}}T^{\frac{4}{3}}\right)\exp\left(-\frac{\pi^{\frac{1}{3}}}{2}X\right) \quad (5.75)$$

for $T,X \geq 0$. For $T = O(1)$ and $X \gg 1$, $F_1(T,X)$ is a steepest descent type integral, and will be dominated in a small neighbourhood of $\bar{\theta} = \pi$ for $X \gg 1$. It is therefore convenient to write

$$\bar{\theta} = \hat{\theta} + \pi$$

in (5.47) to obtain

$$F_1(T,X) = -\int_{-\pi}^{\pi} g_1\left(\hat{\theta} + \pi, T\right)\exp\left(iX\hat{f}_1\left(\hat{\theta} + \pi\right)\right)\exp\left(i\hat{\theta}\right) d\hat{\theta}.$$

for $T, X \geq 0$, with

$$g_1(\hat{\theta} + \pi, T) = \exp\left(-\frac{1}{2\pi^{\frac{1}{3}}} \exp\left(i\frac{\hat{\theta}}{2}\right) T^{\frac{4}{3}}\right)$$

and

$$\hat{f}_1(\hat{\theta} + \pi) = \frac{\pi^{\frac{1}{3}}}{2}\left(-2i \exp\left(-i\frac{\hat{\theta}}{2}\right) - i \exp(i\hat{\theta})\right)$$

so that $F_1(T, X)$ is dominated in a small region about $\hat{\theta} = 0$ for $X \gg 1$. For $\hat{\theta} \ll 1$, we have

$$g_1(\hat{\theta} + \pi, T) = \exp\left(-\frac{T^{\frac{4}{3}}}{2\pi^{\frac{1}{3}}}\left(1 + i\frac{\hat{\theta}}{2} - \frac{\hat{\theta}^2}{8} + O(\hat{\theta}^3)\right)\right) \tag{5.76}$$

$$\exp\left(iX\hat{f}_1(\hat{\theta} + \pi)\right) = \exp\left(\frac{\pi^{\frac{1}{3}}}{2}X\left(3 - \frac{3}{4}\hat{\theta}^2 + O(\hat{\theta}^3)\right)\right). \tag{5.77}$$

We now apply the method of steepest descent to approximate $F_1(T, X)$ for $T = O(1)$ and $X \gg 1$. Thus, via (5.76), (5.77) and using the substitution

$$u^2 = \frac{3\pi^{\frac{1}{3}}}{8}X\hat{\theta}^2$$

we may approximate $F_1(T, X)$, with $T = O(1)$ and $X \gg 1$, as

$$F_1(T, X) = -\frac{2\sqrt{2}}{\sqrt{3}\pi^{\frac{1}{6}}} X^{-\frac{1}{2}} \exp\left(-\frac{T^{\frac{4}{3}}}{2\pi^{\frac{1}{3}}}\right) \exp\left(\frac{3\pi^{\frac{1}{3}}}{2}X\right)\left(1 + O\left(\frac{T^{\frac{4}{3}}}{X}\right)\right)$$

$$\times \left(1 + O\left(\frac{1}{X^{\frac{1}{2}}}\right)\right) \int_{-\infty}^{\infty} \exp\left(-i\lambda u - u^2\right) du \tag{5.78}$$

where

$$\lambda = \frac{T^{\frac{4}{3}}}{\sqrt{6\pi X}}.$$

Using the substitution

$$w = u + i\frac{\lambda}{2}$$

in (5.78) we obtain

$$F_1(T, X) = -\frac{2\sqrt{2}}{\sqrt{3}\pi^{\frac{1}{6}}} X^{-\frac{1}{2}} \exp\left(-\frac{T^{\frac{4}{3}}}{2\pi^{\frac{1}{3}}}\right) \exp\left(\frac{3\pi^{\frac{1}{3}}}{2}X - \frac{T^{\frac{8}{3}}}{24\pi X}\right)$$

$$\times \left(1 + O\left(\frac{T^{\frac{4}{3}}}{X}\right)\right)\left(1 + O\left(\frac{1}{X^{\frac{1}{2}}}\right)\right) \int_{-\infty+i\frac{\lambda}{2}}^{\infty+i\frac{\lambda}{2}} \exp\left(-w^2\right) dw,$$

and an application of the Cauchy residue theorem establishes that

$$\int_{-\infty+i\frac{A}{2}}^{\infty+i\frac{A}{2}} \exp\left(-w^2\right) dw = \sqrt{\pi}.$$

Thus, we obtain the approximation

$$F_1(T,X) = -\frac{2\sqrt{2}\pi^{\frac{1}{3}}}{\sqrt{3}} X^{-\frac{1}{2}} \exp\left(-\frac{T^{\frac{4}{3}}}{2\pi^{\frac{1}{3}}}\right) \exp\left(\frac{3\pi^{\frac{1}{3}}}{2}X\left(1+\frac{T^{\frac{8}{3}}}{36\pi^{\frac{4}{3}}X^2}\right)\right)$$

$$\times\left(1+O\left(\frac{T^{\frac{4}{3}}}{X}\right)\right)\left(1+O\left(\frac{1}{X^{\frac{1}{2}}}\right)\right)$$

$$(5.79)$$

for $T = O(1)$ and $X \gg 1$. It then follows, via (5.75) and (5.79), that

$$F_1(T,X)+F_2(T,X) = -\frac{2\sqrt{2}\pi^{\frac{1}{3}}}{\sqrt{3}} X^{-\frac{1}{2}} \exp\left(-\frac{T^{\frac{4}{3}}}{2\pi^{\frac{1}{3}}}\right) \exp\left(\frac{3\pi^{\frac{1}{3}}}{2}X\left(1+\frac{T^{\frac{8}{3}}}{36\pi^{\frac{4}{3}}X^2}\right)\right)$$

$$\times\left(1+O\left(\frac{T^{\frac{4}{3}}}{X}\right)\right)\left(1+O\left(\frac{1}{X^{\frac{1}{2}}}\right)\right)$$

$$(5.80)$$

for $T = O(1)$ and $X \gg 1$. On returning to (5.72), with (5.80), we finally have that,

$$\tilde{I}(x,\hat{t}) = -\frac{\pi^{\frac{2}{3}}}{\sqrt{6}}\frac{\hat{t}^{\frac{1}{3}}}{x^{\frac{1}{12}}} \exp\left(-\frac{\hat{t}^{\frac{4}{3}}}{2\pi^{\frac{1}{3}}}\right) \exp\left(\frac{3\pi^{\frac{1}{3}}}{2}x^{\frac{1}{2}}\hat{t}^{\frac{2}{3}}\left(1+\frac{\hat{t}^{\frac{4}{3}}}{36\pi^{\frac{4}{3}}x}\right)\right)$$

$$\left(1+O\left(\frac{\hat{t}^{\frac{2}{3}}}{x^{\frac{1}{2}}}\right)\right)\left(1+O\left(\frac{1}{\hat{t}^{\frac{1}{3}}x^{\frac{1}{4}}}\right)\right)$$

$$(5.81)$$

with $\hat{t} = O(1)$ as $x \to \infty$. It is instructive to observe that (5.81) remains a uniform approximation for $O(1) \le \hat{t} < O\left(x^{\frac{3}{4}}\right)$ as $x \to \infty$.

5.1.3   *Coordinate Expansion for $I(x,t)$ as $|x| \to \infty$*

We can now obtain the corresponding approximation for $I(x,t)$, via (5.22), (5.23), (5.24), (5.35), (5.69), (5.71) and (5.81). We obtain:

Region I, $0 \le t \le O\left(x^{-\frac{1}{2}}\right)$ as $x \to \infty$. We have,

$$I(x,\hat{t}) = \frac{1}{\pi^{\frac{5}{3}}}\frac{\hat{t}^{\frac{2}{3}}}{x} \exp\left(-\frac{1}{2}\pi x\right)\left(H_1\left(\hat{t}^{\frac{2}{3}}\right)+H_2\left(\hat{t}^{\frac{2}{3}}\right)\right)\left(1+O\left(\frac{\hat{t}^{\frac{4}{3}}}{x}+\frac{1}{x^{\frac{2}{3}}}+\frac{\hat{t}}{x^{\frac{3}{2}}}\right)\right)$$

$$+O\left(\exp\left(-\pi x\left(1-\frac{1}{\sqrt{2\pi}}\frac{\hat{t}}{x^{\frac{3}{2}}}\right)\right)\right)$$

$$(5.82)$$

with $\tilde{t}\left(=x^{\frac{1}{2}}t\right)=O(1)$ as $x\to\infty$. We also have, via (5.60), (5.63) and (5.67), that,

$$H_1\left(\tilde{t}^{\frac{2}{3}}\right)+H_2\left(\tilde{t}^{\frac{2}{3}}\right)=-2\pi^{\frac{2}{3}}\tilde{t}^{\frac{4}{3}}+O\left(\tilde{t}^{\frac{10}{3}}\right) \tag{5.83}$$

with $0<\tilde{t}\ll 1$, whilst,

$$H_1\left(\tilde{t}^{\frac{2}{3}}\right)=-\frac{2\sqrt{2}\pi^{\frac{1}{3}}}{\sqrt{3}}\frac{1}{\tilde{t}^{\frac{1}{3}}}\exp\left(\frac{3\pi^{\frac{1}{3}}}{2}\tilde{t}^{\frac{2}{3}}\right)\left(1+O\left(\frac{1}{\tilde{t}^{\frac{2}{3}}}\right)\right) \tag{5.84}$$

$$H_2\left(\tilde{t}^{\frac{2}{3}}\right)\leq O\left(\exp\left(-\frac{\pi^{\frac{1}{3}}}{2}\tilde{t}^{\frac{2}{3}}\right)\right) \tag{5.85}$$

with $\tilde{t}\gg 1$.

Region II, $O(1)\leq t<O\left(x^{\frac{1}{4}}\right)$ as $x\to\infty$. We have,

$$I(x,t)=-\frac{4}{\sqrt{6}\pi^{\frac{4}{3}}}\frac{t^{\frac{1}{3}}}{x^{\frac{5}{6}}}\exp\left(-\frac{\pi}{2}x\left(1-\frac{3}{\pi^{\frac{2}{3}}}\frac{t^{\frac{2}{3}}}{x^{\frac{2}{3}}}\right)\right)\left(1+O\left(\frac{1}{x^{\frac{1}{3}}t^{\frac{1}{3}}}+\frac{t^{\frac{4}{3}}}{x^{\frac{1}{3}}}+\frac{1}{x^{\frac{1}{3}}}+\frac{t}{x}\right)\right)$$

$$+O\left(\exp\left(-\pi x\left(1-\frac{1}{\sqrt{2\pi}}\frac{t}{x}\right)\right)\right)$$

$$\tag{5.86}$$

with $t=O(1)$ as $x\to\infty$.

Region III, $O\left(x^{\frac{1}{4}}\right)\leq t<O(x)$ as $x\to\infty$. We have,

$$I(x,\hat{t})=-\frac{4}{\sqrt{6}\pi^{\frac{4}{3}}}\frac{\hat{t}^{\frac{1}{3}}}{x^{\frac{3}{4}}}\exp\left(-\frac{\pi}{2}x\left(1-\frac{3}{\pi^{\frac{2}{3}}}\frac{\hat{t}^{\frac{2}{3}}}{x^{\frac{1}{2}}}+\frac{1}{\pi^{\frac{4}{3}}}\frac{\hat{t}^{\frac{4}{3}}}{x}-\frac{1}{12\pi^2}\frac{\hat{t}^2}{x^{\frac{3}{2}}}\right)\right)$$

$$\times\left(1+O\left(\frac{1}{x^{\frac{2}{3}}}+\frac{\hat{t}}{x^{\frac{3}{4}}}+\frac{\hat{t}^{\frac{1}{3}}}{x^{\frac{1}{4}}}+\frac{1}{\hat{t}^{\frac{2}{3}}x^{\frac{1}{2}}}\right)\right)+O\left(\exp\left(-\pi x\left(1-\frac{1}{\sqrt{2\pi}}\frac{\hat{t}}{x^{\frac{3}{4}}}\right)\right)\right)$$

$$\tag{5.87}$$

with $\hat{t}\left(=x^{-\frac{1}{4}}t\right)=O(1)$ as $x\to\infty$.

It is clear that asymptotic matching between each region is satisfied, and it can also be verified that the approximation in Region I asymptotically matches at the leading order with the approximation to $I(x,t)$, for $t\to 0$ with $x\gg 1$, in (4.17).

5.2 Coordinate Expansion for $\bar{\eta}(x,t)$ as $|x| \to \infty$

We can now construct the approximation for $\bar{\eta}(x,t)$ in each region. We have, via (3.32) and (3.34),

$$\bar{\eta}(x,t) = \frac{1}{2\pi\beta}\left(I(x,t) - I(x-\beta,t)\right) \tag{5.88}$$

for $x \geq 0$, $t \geq 0$, and

$$\bar{\eta}(x,t) = \frac{1}{2\pi\beta}\left(2\pi\beta + I(-x,t) - I(\beta-x,t)\right) \tag{5.89}$$

for $x \leq 0$, $t \geq 0$. In order to construct $\bar{\eta}(x,t)$ we must first consider the approximation for $I(x-\beta,t)$ in each region. In Region I, the approximation of $I(x-\beta,t)$ requires the scaled coordinate $\tilde{t}' = (x-\beta)^{\frac{1}{2}}t$ as $x \to \infty$, from which we have $\tilde{t}' = \tilde{t}\left(1-\frac{\beta}{x}\right)^{\frac{1}{2}}$. Thus, in Region I we have

$$\bar{\eta}(x,\tilde{t}) = \frac{1}{2\pi\beta}\left(I(x,\tilde{t}) - I\left(x-\beta,\tilde{t}\left(1-\frac{\beta}{x}\right)^{\frac{1}{2}}\right)\right) \tag{5.90}$$

with $\tilde{t}\left(=x^{\frac{1}{2}}t\right) = O(1)$ as $x \to \infty$. Similarly, in Region III the approximation of $I(x-\beta,t)$ requires the scaled coordinate $\hat{t}' = (x-\beta)^{-\frac{1}{4}}t$ as $x \to \infty$, from which we have $\hat{t}' = \hat{t}\left(1-\frac{\beta}{x}\right)^{-\frac{1}{4}}$. Thus, in Region III we have

$$\bar{\eta}(x,\hat{t}) = \frac{1}{2\pi\beta}\left(I(x,\hat{t}) - I\left(x-\beta,\hat{t}\left(1-\frac{\beta}{x}\right)^{-\frac{1}{4}}\right)\right) \tag{5.91}$$

with $\hat{t}\left(=x^{-\frac{1}{4}}t\right) = O(1)$ as $x \to \infty$. Thus, via (5.82), (5.86), (5.87), (5.88), (5.89), (5.90) and (5.91), we have the approximation for $\bar{\eta}(x,t)$ in the following regions,

Region I$^+$, $0 \leq t \leq O\left(x^{-\frac{1}{2}}\right)$ as $x \to +\infty$. We have,

$$\bar{\eta}(x,\tilde{t}) = \frac{1}{2\pi\beta}\left(\frac{1}{\pi^{\frac{5}{3}}}\frac{\tilde{t}^{\frac{2}{3}}}{x}\exp\left(-\frac{1}{2}\pi x\right)\right)\left(H_1\left(\tilde{t}^{\frac{2}{3}}\right) + H_2\left(\tilde{t}^{\frac{2}{3}}\right)\right.$$

$$-\exp\left(\frac{1}{2}\pi\beta\right)\left(H_1\left(\tilde{t}^{\frac{2}{3}}\left(1-\frac{\beta}{x}\right)^{\frac{1}{3}}\right) + H_2\left(\tilde{t}^{\frac{2}{3}}\left(1-\frac{\beta}{x}\right)^{\frac{1}{3}}\right)\right)\right)$$

$$\times\left(1 + O\left(\frac{\tilde{t}^{\frac{4}{3}}}{x} + \frac{1}{x^{\frac{2}{3}}} + \frac{\tilde{t}}{x^{\frac{3}{2}}}\right)\right) + O\left(\exp\left(-\pi x\left(1-\frac{1}{\sqrt{2\pi}}\frac{\tilde{t}}{x^{\frac{3}{2}}}\right)\right)\right)$$

with $\tilde{t} = x^{\frac{1}{2}}t = O(1)$ as $x \to \infty$, and the approximations for $H_1\left(\tilde{t}^{\frac{2}{3}}\right)$ and $H_2\left(\tilde{t}^{\frac{2}{3}}\right)$ given in (5.83), (5.84) and (5.85).

Region II$^+$, $O(1) \leq t < O\left(x^{\frac{1}{4}}\right)$ as $x \to +\infty$. We have,

$$\bar{\eta}(x,t) = \frac{1}{2\pi\beta}\left(\exp\left(\frac{\pi}{2}\beta\right) - 1\right)\left(\frac{4}{\sqrt{6}\pi^{\frac{4}{3}}}\frac{t^{\frac{1}{3}}}{x^{\frac{5}{6}}}\exp\left(-\frac{\pi}{2}x\left(1 - \frac{3}{\pi^{\frac{2}{3}}}\frac{t^{\frac{2}{3}}}{x^{\frac{2}{3}}}\right)\right)\right)$$

$$\times\left(1 + O\left(\frac{1}{x^{\frac{1}{3}}t^{\frac{2}{3}}} + \frac{t^{\frac{4}{3}}}{x^{\frac{1}{3}}} + \frac{1}{x^{\frac{2}{3}}} + \frac{t}{x} + \frac{t^{\frac{2}{3}}}{x^{\frac{2}{3}}}\right)\right) + O\left(\exp\left(-\pi x\left(1 - \frac{1}{\sqrt{2\pi}}\frac{t}{x}\right)\right)\right)$$

with $t = O(1)$ as $x \to \infty$.

Region III$^+$, $O\left(x^{\frac{1}{4}}\right) \leq t < O(x)$ as $x \to +\infty$. We have,

$$\bar{\eta}(x,\hat{t}) = \frac{1}{2\pi\beta}\left(\exp\left(\frac{\pi}{2}\beta\right) - 1\right)\left(\frac{4}{\sqrt{6}\pi^{\frac{4}{3}}}\frac{\hat{t}^{\frac{1}{3}}}{x^{\frac{3}{4}}}\exp\left(-\frac{\pi}{2}x\left(1 - \frac{3}{\pi^{\frac{2}{3}}}\frac{\hat{t}^{\frac{2}{3}}}{x^{\frac{1}{2}}} + \frac{1}{\pi^{\frac{4}{3}}}\frac{\hat{t}^{\frac{4}{3}}}{x}\right.\right.\right.$$

$$\left.\left.\left. - \frac{1}{12\pi^2}\frac{\hat{t}^2}{x^{\frac{3}{2}}}\right)\right)\right)\times\left(1 + O\left(\frac{1}{x^{\frac{2}{3}}} + \frac{\hat{t}}{x^{\frac{3}{4}}} + \frac{\hat{t}^{\frac{2}{3}}}{x^{\frac{1}{2}}} + \frac{1}{\hat{t}^{\frac{1}{3}}x^{\frac{1}{4}}} + \frac{\hat{t}^{\frac{4}{3}}}{x} + \frac{\hat{t}^2}{x^{\frac{3}{2}}}\right)\right)$$

$$+ O\left(\exp\left(-\pi x\left(1 - \frac{1}{\sqrt{2\pi}}\frac{\hat{t}}{x^{\frac{3}{4}}}\right)\right)\right)$$

$$(5.92)$$

with $\hat{t}\left(= x^{-\frac{1}{4}}t\right) = O(1)$ as $x \to \infty$.

Region I$^-$, $0 \leq t \leq O\left((-x)^{-\frac{1}{2}}\right)$ as $x \to -\infty$. We have,

$$\bar{\eta}(x,\hat{t}) = 1 + \frac{1}{2\pi\beta}\left(\frac{1}{\pi^{\frac{5}{3}}}\frac{\tilde{t}^{\frac{2}{3}}}{(-x)}\exp\left(\frac{1}{2}\pi x\right)\right)\left(H_1\left(\tilde{t}^{\frac{2}{3}}\right) + H_2\left(\tilde{t}^{\frac{2}{3}}\right)\right.$$

$$\left. - \exp\left(-\frac{1}{2}\pi\beta\right)\left(H_1\left(\tilde{t}^{\frac{2}{3}}\left(1 - \frac{\beta}{x}\right)^{\frac{1}{3}}\right) + H_2\left(\tilde{t}^{\frac{2}{3}}\left(1 - \frac{\beta}{x}\right)^{\frac{1}{3}}\right)\right)\right)$$

$$\times\left(1 + O\left(\frac{\tilde{t}^{\frac{4}{3}}}{(-x)} + \frac{1}{(-x)^{\frac{2}{3}}} + \frac{\tilde{t}}{(-x)^{\frac{3}{2}}}\right)\right) + O\left(\exp\left(\pi x\left(1 - \frac{1}{\sqrt{2\pi}}\frac{\tilde{t}}{(-x)^{\frac{3}{2}}}\right)\right)\right)$$

with $\tilde{t}\left(= (-x)^{\frac{1}{2}}t\right) = O(1)$ as $x \to -\infty$, with the approximations for $H_1\left(\tilde{t}^{\frac{2}{3}}\right)$ and $H_2\left(\tilde{t}^{\frac{2}{3}}\right)$ given in (5.83), (5.84) and (5.85).

Region II⁻, $O(1) \leq t < O\left(x^{\frac{1}{4}}\right)$ as $x \to -\infty$. We have,

$$\bar{\eta}(x,t) = 1 - \frac{1}{2\pi\beta}\left(1 - \exp\left(-\frac{\pi}{2}\beta\right)\right)\left(\frac{4}{\sqrt{6}\pi^{\frac{4}{3}}}\frac{t^{\frac{1}{3}}}{(-x)^{\frac{5}{6}}}\right.$$

$$\times \exp\left(\frac{\pi}{2}x\left(1 - \frac{3}{\pi^{\frac{2}{3}}}\frac{t^{\frac{2}{3}}}{(-x)^{\frac{2}{3}}}\right)\right)\right)\left(1 + O\left(\frac{1}{(-x)^{\frac{1}{3}}t^{\frac{2}{3}}} + \frac{t^{\frac{4}{3}}}{(-x)^{\frac{1}{3}}} + \frac{1}{(-x)^{\frac{2}{3}}}\right.\right.$$

$$\left.\left.+ \frac{t}{(-x)} + \frac{t^{\frac{2}{3}}}{(-x)^{\frac{2}{3}}}\right)\right) + O\left(\exp\left(\pi x\left(1 - \frac{1}{\sqrt{2\pi}}\frac{t}{(-x)}\right)\right)\right)$$

with $t = O(1)$ as $x \to -\infty$.

Region III⁻, $O\left(x^{\frac{1}{4}}\right) \leq t < O(x)$ as $x \to -\infty$. We have,

$$\bar{\eta}(x,\hat{t}) = 1 - \frac{1}{2\pi\beta}\left(1 - \exp\left(-\frac{\pi}{2}\beta\right)\right)\left(\frac{4}{\sqrt{6}\pi^{\frac{4}{3}}}\frac{\hat{t}^{\frac{1}{3}}}{(-x)^{\frac{3}{4}}}\exp\left(\frac{\pi}{2}x\left(1 - \frac{3}{\pi^{\frac{2}{3}}}\frac{\hat{t}^{\frac{2}{3}}}{(-x)^{\frac{1}{2}}}\right.\right.\right.$$

$$\left.\left.\left.+ \frac{1}{\pi^{\frac{4}{3}}}\frac{\hat{t}^{\frac{4}{3}}}{(-x)} - \frac{1}{12\pi^2}\frac{\hat{t}^2}{(-x)^{\frac{3}{2}}}\right)\right)\right)\times\left(1 + O\left(\frac{1}{(-x)^{\frac{2}{3}}} + \frac{\hat{t}}{(-x)^{\frac{3}{4}}} + \frac{\hat{t}^{\frac{2}{3}}}{(-x)^{\frac{1}{2}}}\right.\right.$$

$$\left.\left.+ \frac{1}{\hat{t}^{\frac{1}{3}}(-x)^{\frac{1}{4}}} + \frac{\hat{t}^{\frac{4}{3}}}{(-x)} + \frac{\hat{t}^2}{(-x)^{\frac{3}{2}}}\right)\right) + O\left(\exp\left(\pi x\left(1 - \frac{1}{\sqrt{2\pi}}\frac{\hat{t}}{(-x)^{\frac{3}{4}}}\right)\right)\right)$$

$$\tag{5.93}$$

with $\hat{t}\left(= (-x)^{-\frac{1}{4}}t\right) = O(1)$ as $x \to -\infty$.

We now move onto consider coordinate expansions for $\bar{\eta}(x,t)$ as $t \to \infty$.

Chapter 6

Coordinate Expansions for $\bar{\eta}(x,t)$ as $t \to \infty$

In this chapter we consider $\bar{\eta}(x,t)$, as given in (3.30), as $t \to \infty$ with $x \in \mathbb{R}$. The natural spatial coordinate as $t \to \infty$ is $y = \left(\frac{x}{t}\right)$. We now construct an approximation to $\bar{\eta}(y,t)$ as $t \to \infty$, uniformly for $y \in \mathbb{R}$. We find that the approximation consists of four outer regions, two of which exhibit oscillatory behaviour and two that show an exponentially small disturbance in the far field conditions. We then find inner region approximations for $\bar{\eta}(y,t)$, which connect the oscillatory regions to the exponentially decaying regions, which are in terms of Airy functions and their integrals and connect the oscillatory regions to the exponentially decaying regions.

6.1 Outer Region Coordinate Expansion for $\bar{\eta}(x,t)$ as $t \to \infty$

We examine $\bar{\eta}(x,t)$, as given in (3.30), as $t \to \infty$ with $x \in \mathbb{R}$. We first approximate $\bar{\eta}(x,t)$ for $x = O(t)$ as $t \to \infty$. Introducing $y = \frac{x}{t}$, we write via (3.30),

$$\bar{\eta}(y,t) = \frac{1}{2\pi\beta} \int_{C_\delta} f(k) \cos\left(\gamma(k)t\right) \exp\left(-ikty\right) dk$$
$$- \frac{i}{2\pi} \int_{C_\delta} \frac{1}{k} \cos\left(\gamma(k)t\right) \exp\left(-ikty\right) dk \qquad (6.1)$$

for $(y,t) \in \mathbb{R} \times \mathbb{R}^+$, where

$$f(k) = \frac{1}{k^2}\left(1 - \exp\left(i\beta k\right)\right) + i\frac{\beta}{k}$$

with $k \in \mathbb{C}$, and the contour C_δ shown in Fig. (6.1), where $C_\delta = L_1 \cup L_2 \cup C^\delta$. We now consider the first integral on the right-hand side of (6.1). Observing that $f(k)$

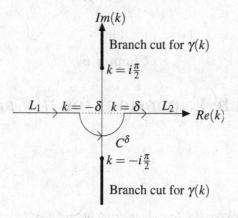

Fig. 6.1: Contour C_δ in the complex k-plane.

is entire, Cauchy's Theorem gives

$$\int_{C_\delta} f(k) \cos\left(\gamma(k)t\right) \exp\left(-ikty\right) dk = \int_{-\infty}^{\infty} f(k) \cos\left(\gamma(k)t\right) \exp\left(-ikty\right) dk$$

$$= \frac{1}{2} \int_{-\infty}^{\infty} f(k) \exp\left(it\left(\gamma(k) - ky\right)\right) dk \qquad (6.2)$$

$$+ \frac{1}{2} \int_{-\infty}^{\infty} f(k) \exp\left(-it\left(\gamma(k) + ky\right)\right) dk$$

with $(y,t) \in \mathbb{R} \times \mathbb{R}^+$. We now consider the second integral on the right-hand side of (6.1). Along L_1 and L_2 we have $-\infty < k \leq -\delta$ and $\delta \leq k < \infty$ respectively, and on C^δ we may write

$$k = \delta \exp\left(i\theta\right),$$

for $\theta \in [-\pi, 0]$, so that, with $0 < \delta \ll 1$, we have

$$\int_{C_\delta} \frac{1}{k} \cos\left(\gamma(k)t\right) \exp\left(-ikty\right) dk = \int_{-\infty}^{-\delta} \frac{1}{k} \cos\left(\gamma(k)t\right) \exp\left(-ikty\right) dk$$

$$+ \int_{\delta}^{\infty} \frac{1}{k} \cos\left(\gamma(k)t\right) \exp\left(-ikty\right) dk$$

$$+ i \int_{-\pi}^{0} \left(1 + O\left(\delta\right)\right) d\theta$$

with $(y,t) \in \mathbb{R} \times \mathbb{R}^+$. Thus,

$$\int_{C_\delta} \frac{1}{k} \cos\left(\gamma(k)t\right) \exp\left(-ikty\right) dk = i\pi - 2i \int_\delta^\infty \frac{1}{k} \cos\left(\gamma(k)t\right) \sin\left(kty\right) dk + O(\delta) \tag{6.3}$$

for $(y,t) \in \mathbb{R} \times \mathbb{R}^+$. Now, the left-hand side of (6.3) is independent of $0 < \delta < \frac{\pi}{2}$ (via Cauchy's Theorem) and so, after using the identity

$$2\cos\left(\gamma(k)t\right)\sin\left(kty\right) = \sin t\left(\gamma(k) + ky\right) - \sin t\left(\gamma(k) - ky\right),$$

and taking $\delta \to 0$ on the right-hand side of (6.3), we obtain

$$\int_{C_\delta} \frac{1}{k} \cos\left(\gamma(k)t\right) \exp\left(-ikty\right) dk = i\pi + i \int_0^\infty \frac{1}{k} \sin t\left(\gamma(k) - ky\right) dk$$
$$- i \int_0^\infty \frac{1}{k} \sin t\left(\gamma(k) + ky\right) dk \tag{6.4}$$

for $(y,t) \in \mathbb{R} \times \mathbb{R}^+$. Finally, via (6.1), (6.2) and (6.4), we have

$$\bar{\eta}(y,t) = \frac{1}{2} + \frac{1}{4\pi\beta} \left(J_+(y,t) + J_-(y,t)\right) + \frac{1}{2\pi} \left(F(y,t) - F(-y,t)\right) \tag{6.5}$$

for $(y,t) \in \mathbb{R} \times \mathbb{R}^+$, where

$$J_+(y,t) = \int_{-\infty}^\infty f(k) \exp\left(it\left(\gamma(k) - ky\right)\right) dk, \tag{6.6}$$

$$J_-(y,t) = \int_{-\infty}^\infty f(k) \exp\left(-it\left(\gamma(k) + ky\right)\right) dk, \tag{6.7}$$

$$F(y,t) = \int_0^\infty \frac{1}{k} \sin t\left(\gamma(k) - ky\right) dk, \tag{6.8}$$

and

$$f(k) = \frac{1}{k^2} \left(1 - \exp\left(i\beta k\right)\right) + i\frac{\beta}{k} \tag{6.9}$$

for $k \in \mathbb{C}$, which is entire. The integrals in (6.6)–(6.8) are now in a suitable form to be estimated.

6.2   Outer Region Coordinate Expansions for $J_+(y,t)$ and $J_-(y,t)$ as $t \to \infty$

We first consider $J_+(y,t)$ in (6.6) as $t \to \infty$ with $y = O(1)$. We observe that $J_+(y,t)$ is in the form of a stationary phase integral. There are three distinct asymptotic regions for $J_+(y,t)$ as $t \to \infty$, namely.

Region I: $o(1) < y < 1 - o(1)$ as $t \to \infty$,

Region II: $y < -o(1)$ as $t \to \infty$,

Region III: $y > 1 + o(1)$ as $t \to \infty$.

We consider the approximation of $J_+(y,t)$ in each region in turn. We begin in Region I. The phase becomes stationary in (6.6) at those values $k \in \mathbb{R}$ when

$$\gamma'(k) = y. \tag{6.10}$$

Recall that

$$\gamma(k) = (k \tanh k)^{\frac{1}{2}} \tag{6.11}$$

for $-\infty < k < \infty$, with the branch defined so that

$$\gamma(-k) = -\gamma(k)$$

with $k > 0$. Graphs of $\gamma(k)$ and $\gamma'(k)$ are given in Figs. (6.2) and (6.3). It follows from (6.10) that $J_+(y,t)$ has exactly two points of stationary phase when $o(1) < y < 1 - o(1)$, and we denote these points to be at

$$k = \pm k_s(y) \tag{6.12}$$

for $o(1) < y < 1 - o(1)$, where $k_s(y) > 0$. At this stage it is useful to examine the structure of $k_s(y)$. As $y \to 1^-$, we have $k_s(y) \to 0$. Thus, via (6.10), we may write

$$\gamma'(0) + \gamma''(0)k_s(y) + \frac{1}{2}\gamma'''(0)k_s(y)^2 + O\left(k_s(y)^4\right) = y$$

as $y \to 1^-$. Hence

$$k_s(y) = \sqrt{2}(1-y)^{\frac{1}{2}} + O\left((1-y)^{\frac{3}{2}}\right) \tag{6.13}$$

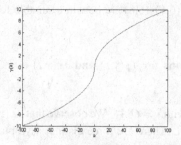

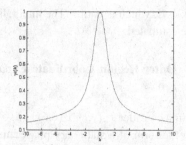

Fig. 6.2: The graph of $\gamma(k)$.   Fig. 6.3: The graph of $\gamma'(k)$.

as $y \to 1^-$, where $\gamma'''(0) = -1$. As $y \to 0^+$, we have $k_s(y) \to \infty$. Thus, via (6.11) we have

$$\gamma'(k_s(y)) = \frac{1}{2}k_s(y)^{-\frac{1}{2}} + O\left(k_s(y)^{\frac{1}{2}}\exp\left(-2k_s(y)\right)\right) \tag{6.14}$$

as $y \to 0^+$. Hence

$$k_s(y) = \frac{1}{4y^2} + O\left(\frac{1}{y^4}\exp\left(\frac{-1}{2y^2}\right)\right) \tag{6.15}$$

as $y \to 0^+$. Using (6.13) and (6.15), and observing that $k_s(y)$ is decreasing with $y \in (0,1)$, we give a sketch of $k_s(y)$ in Fig. (6.4). It is therefore convenient to write (6.6) as

$$J_+(y,t) = K_1(y,t) + K_2(y,t) \tag{6.16}$$

where

$$K_1(y,t) = \int_{-\infty}^{0} f(k)\exp\left(it\left(\gamma(k) - ky\right)\right) dk, \tag{6.17}$$

$$K_2(y,t) = \int_{0}^{\infty} f(k)\exp\left(it\left(\gamma(k) - ky\right)\right) dk,$$

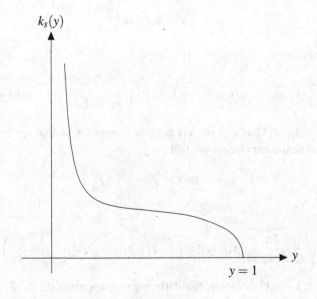

Fig. 6.4: A sketch of $k_s(y)$ for $y \in (0,1)$.

and each integral has one point of stationary phase. We first consider $K_2(y,t)$. The phase is stationary at $k = k_s(y)$, so we may approximate $K_2(y,t)$ as

$$K_2(y,t) = \exp\left(-itk_s(y)y\right) \int_{-\delta(t)}^{\delta(t)} f(w+k_s(y)) \exp\left(it\left(\gamma(w+k_s(y)) - wy\right)\right) dw$$

(6.18)

for $o(1) < y < 1 - o(1)$ as $t \to \infty$, where $\delta(t) = o(1)$ as $t \to \infty$ and we have used the substitution $w = k - k_s(y)$. For $|w| \ll 1$ we have

$$f(w+k_s(y)) = f(k_s(y)) + f'(k_s(y))w + \frac{1}{2}f''(k_s(y))w^2 + O\left(f'''(k_s(y))w^3\right)$$

$$\gamma(w+k_s(y)) = \gamma(k_s(y)) + \gamma'(k_s(y))w + \frac{1}{2}\gamma''(k_s(y))w^2 + O\left(\gamma'''(k_s(y))w^3\right)$$

from which (6.18) becomes

$$K_2(y,t) \frown f(k_s(y)) \exp\left(-it\left(k_s(y)y - \gamma(k_s(y))\right)\right) \int_{-\delta(t)}^{\delta(t)} \exp\left(i\frac{t}{2}\gamma''(k_s(y))w^2\right) dw$$

(6.19)

for $o(1) < y < 1 - o(1)$ as $t \to \infty$. As $\gamma''(k) < 0$ for $k > 0$, and $\gamma''(-k) = -\gamma''(k)$, we use the substitution

$$u^2 = \frac{t}{2}\gamma''(-k_s(y))w^2$$

(6.20)

in (6.19) to obtain, on taking $\delta(t) = t^{-\frac{2}{5}}$,

$$K_2(y,t) \frown f(k_s(y)) \exp\left(-it\left(k_s(y)y - \gamma(k_s(y))\right)\right) \sqrt{\frac{2}{t\gamma''(-k_s(y))}} \int_{-\infty}^{\infty} \exp\left(-iu^2\right) du$$

for $o(1) < y < 1 - o(1)$ as $t \to \infty$. Via contour integration and an application of Cauchy's Theorem it may be established that

$$\int_{-\infty}^{\infty} \exp\left(-iu^2\right) = \sqrt{\pi}\exp\left(-i\frac{\pi}{4}\right)$$

(6.21)

and so

$$K_2(y,t) \frown \sqrt{\frac{2\pi}{t\gamma''(-k_s(y))}} f(k_s(y)) \exp\left(-i\left(t\left(k_s(y)y - \gamma(k_s(y))\right) + \frac{\pi}{4}\right)\right) \quad (6.22)$$

for $o(1) < y < 1 - o(1)$ as $t \to \infty$. Similarly, we may approximate (6.17) as

$$K_1(y,t) \frown \sqrt{\frac{2\pi}{t\gamma''(-k_s(y))}} f(-k_s(y)) \exp\left(i\left(t\left(k_s(y)y - \gamma(k_s(y))\right) + \frac{\pi}{4}\right)\right) \quad (6.23)$$

for $o(1) < y < 1 - o(1)$ as $t \to \infty$. Therefore, via (6.9), (6.16), (6.22) and (6.23), we obtain

$$
\begin{aligned}
J_+(y,t) \sim & \sqrt{\frac{2\pi}{t\gamma''(-k_s(y))}} \left(\frac{2\beta}{k_s(y)} \sin\left(\frac{\pi}{4} + tk_s(y)\, (y - c(k_s(y))) \right) \right. \\
& + \frac{2}{k_s(y)^2} \left(\cos\left(\frac{\pi}{4} + tk_s(y)\, (y - c(k_s(y))) \right) \right. \\
& \left. \left. - \cos\left(\frac{\pi}{4} - \beta k_s(y) + tk_s(y)\, (y - c(k_s(y))) \right) \right) \right)
\end{aligned}
\tag{6.24}
$$

for $o(1) < y < 1 - o(1)$ as $t \to \infty$, where

$$
c(k) = \frac{\gamma(k)}{k}
$$

for $k \geq 0$. It is now instructive to consider the forms of (6.24) when $0 < y \ll 1$ and $0 < 1 - y \ll 1$. These are readily obtained, via (6.11), (6.13) and (6.14), as

$$
\begin{aligned}
J_+(y,t) \sim & \frac{8\sqrt{\pi} y^{\frac{1}{2}}}{t^{\frac{1}{2}}} \left(\beta \sin\left(\frac{\pi}{4} - \frac{t}{4y} \right) + 4y^2 \left(\cos\left(\frac{\pi}{4} - \frac{t}{4y} \right) \right. \right. \\
& \left. \left. - \cos\left(\frac{\pi}{4} - \frac{\beta}{4y^2} - \frac{t}{4y} \right) \right) \right)
\end{aligned}
\tag{6.25}
$$

with $0 < y \ll 1$ as $t \to \infty$, and

$$
J_+(y,t) \sim \frac{2^{\frac{1}{4}} \beta^2 \sqrt{\pi}}{t^{\frac{1}{2}} (1-y)^{\frac{1}{4}}} \cos\left(\frac{\pi}{4} - \frac{2\sqrt{2}}{3} t(1-y)^{\frac{3}{2}} \right)
\tag{6.26}
$$

with $0 < (1 - y) \ll 1$ as $t \to \infty$.

We now consider the approximation of $J_+(y,t)$ in Region III. Via (6.6), we write

$$
J_+(y,t) = \int_{-\infty}^{\infty} f(k) \exp\left(itg(k,y) \right) dk,
\tag{6.27}
$$

where

$$
g(k,y) = \gamma(k) - yk
\tag{6.28}
$$

for $y > 1$ with $t \in \mathbb{R}^+$. With $y > 1$, (6.27) is a steepest descents integral as $t \to \infty$. In particular, $g(k,y)$ becomes stationary in (6.27) at those values $k = i\tau$, with $\tau \in \left(-\frac{\pi}{2}, \frac{\pi}{2} \right)$, when

$$
g_k(i\tau, y) = 0.
\tag{6.29}
$$

It follows from (6.28) and (6.29) that for $y > 1$, $g(k,y)$ has two stationary points on $k = i\tau$, with $\tau \in \left(-\frac{\pi}{2}, \frac{\pi}{2} \right)$, and we denote these points as

$$
k = \pm i\tau_s(y)
\tag{6.30}
$$

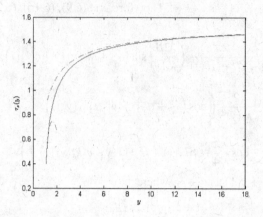

Fig. 6.5: The graph of $\tau_s(y)$ for $y > 1$, with the asymptotic approximations $(--)$ from (6.57).

where $\tau_s(y) \in \left(0, \frac{\pi}{2}\right)$ for $y > 1$. A graph of $\tau_s(y)$ is given in Fig. (6.5). Setting $k = \sigma + i\tau$, with $\sigma, \tau \in \mathbb{R}$, we may write (6.28) as

$$g(k,y) = \gamma(k) - yk = u(k,y) + iv(k,y), \tag{6.31}$$

with $y > 1$, where $u(k,y)$ and $v(k,y)$ denote the real and imaginary parts of $g(k,y)$ respectively. Via (3.23), (3.24), (3.25) and (6.31), we may determine the qualitative behaviour of the level curves for $u(k,y)$ and $v(k,y)$, and sketches of these level curves are given in Figs. (6.7) and (6.6) respectively, where the arrows point in the direction of increasing $v(k,y)$ and $u(k,y)$. We observe that the level curve D, in Fig. (6.7), is the path of steepest descent for $g(k,y)$ in (6.27). Thus we will deform the contour in (6.27) onto the steepest descent contour D. On D, $k = k_d(\sigma) = \sigma + i\tau_d(\sigma)$, with $\tau_d(0) = -\tau_s(y)$ and $\tau_d(-\sigma) = \tau_d(\sigma)$, where $\tau_d(\sigma)$ is monotone decreasing for $\sigma \geq 0$. Moreover,

$$\tau_d(\sigma) \sim -2y^2\sigma^2 \quad \text{as } |\sigma| \to \infty. \tag{6.32}$$

We now consider the contour $C_L = [-L, L] \cup L_1 \cup L_2 \cup D_L$, where L_1 and L_2 are arcs on the circle $|k| = L$, and D_L is a finite section of D, as shown in Fig. (6.8). The points $k = k_1 = \sigma_1 + i\tau_1$ and $k = k_2 = \sigma_2 + i\tau_2$ are the intersection points of L_1 and L_2 with D_L respectively. With $L \gg 1$, we have, via (6.32), that

$$\sigma_1 = \frac{L^{\frac{1}{2}}}{\sqrt{2}y}\left(1 - \frac{1}{8y^2L} + O\left(\frac{1}{y^4L^2}\right)\right) \tag{6.33}$$

$$\tau_1 = -L\left(1 - \frac{1}{4y^2L} + O\left(\frac{1}{y^4L^2}\right)\right) \tag{6.34}$$

as $L \to \infty$. With θ_L being the angle shown in Fig. (6.8), we also have from (6.33) and (6.34) that

$$\theta_L = \frac{\pi}{2} - \frac{1}{\sqrt{2}yL^{\frac{1}{2}}} + O\left(\frac{1}{y^3 L^{\frac{3}{2}}}\right) \tag{6.35}$$

as $L \to \infty$. Now, via Cauchy's Theorem, we have

$$\int_{C_L} f(k) \exp\left(itg(k,y)\right) dk = 0, \tag{6.36}$$

where

$$f(k) = \frac{1}{k^2}\left(1 - \exp\left(i\beta k\right)\right) + i\frac{\beta}{k}. \tag{6.37}$$

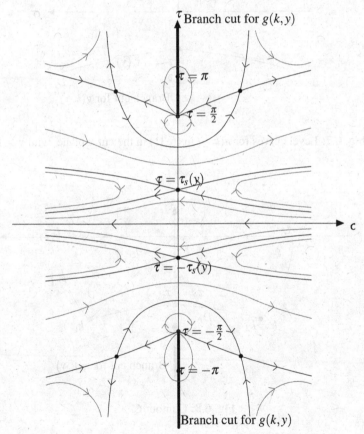

Fig. 6.6: Level curves for $v(k,y)$ in (6.31) in the cut k-plane, with $y > 1$.

and

$$g(k,y) = \gamma(k) - yk = u(k,y) + iv(k,y), \qquad (6.38)$$

for $y > 1$ with $t \in \mathbb{R}^+$. Therefore

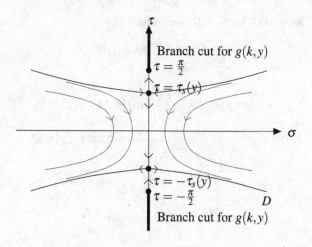

Fig. 6.7: Level curves for $u(k,y)$ in (6.31) in the cut k-plane, with $y > 1$.

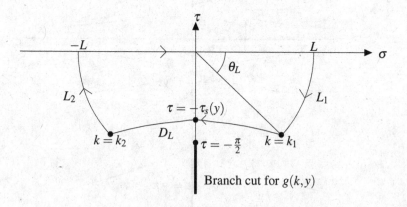

Fig. 6.8: Contour C_L.

$$\int_{C_L} f(k)\exp\left(itg(k,y)\right) dk = \int_{-L}^{L} f(k)\exp\left(itg(k,y)\right) dk + \int_{L_1} f(k)\exp\left(itg(k,y)\right) dk$$

$$+ \int_{D_L} f(k)\exp\left(itg(k,y)\right) dk + \int_{L_2} f(k)\exp\left(itg(k,y)\right) dk$$

$$= 0 \tag{6.39}$$

for $y > 1$ with $t \in \mathbb{R}^+$. We now consider the integral along L_1 in (6.39). Set

$$J_{L_1}(y,t) = \int_{L_1} f(k)\exp\left(itg(k,y)\right) dk, \tag{6.40}$$

for $y > 1$ with $t \in \mathbb{R}^+$. On L_1 we write

$$k = L(\cos\theta + i\sin\theta) \tag{6.41}$$

for $\theta \in [-\theta_L, 0]$. Via (6.35), (6.37), (6.38), (6.40) and (6.41) we have

$$|J_{L_1}(y,t)| \le \int_{-\frac{\pi}{2}}^{0} \left(\frac{1}{L}(1 + \exp(-\beta L\sin\theta)) + \beta\right)$$
$$\times \exp\left(t\left(yL\sin\theta - 2L^{\frac{1}{2}}\sin\frac{\theta}{2}\right)\right) d\theta \tag{6.42}$$

for $y > 1$ with $t \in \mathbb{R}^+$. It is readily established that

$$\int_{-\frac{\pi}{2}}^{0} \exp\left(t\left(yL\sin\theta - 2L^{\frac{1}{2}}\sin\frac{\theta}{2}\right)\right) d\theta \le \frac{16}{L} \tag{6.43}$$

and

$$\int_{-\frac{\pi}{2}}^{0} \exp\left(t\left(yL\sin\theta - 2L^{\frac{1}{2}}\sin\frac{\theta}{2}\right) - \beta L\sin\theta\right) d\theta \le \frac{16}{L} \tag{6.44}$$

for $y > 1$ with $t > \beta + 1$ and L sufficiently large. Thus, via (6.42), (6.43) and (6.44), we have

$$J_{L_1}(y,t) = \int_{L_1} f(k)\exp\left(itg(k,y)\right) dk \to 0 \quad \text{as } L \to \infty \tag{6.45}$$

for $y > 1$ with $t > \beta + 1$. Similarly, for the integral along L_2 in (6.39), we have

$$J_{L_2}(y,t) = \int_{L_2} f(k)\exp\left(itg(k,y)\right) dk \to 0 \quad \text{as } L \to \infty \tag{6.46}$$

for $y > 1$ with $t > \beta + 1$. It now follows from (6.27), (6.39), (6.45) and (6.46) that, on letting $L \to \infty$, we have

$$J_+(y,t) = -\int_{D} f(k)\exp\left(itg(k,y)\right) dk \tag{6.47}$$

for $y > 1$ with $t > \beta + 1$, whilst $f(k)$ and $g(k,y)$ are given by (6.37) and (6.38), where $u(k,y)$ and $v(k,y)$ are the real and imaginary parts of $g(k,y)$ respectively, and we recall that

$$u(k) = 0 \text{ on } D. \tag{6.48}$$

The level curve D is the path of steepest descent for $g(k,y)$ in (6.38), and it is observed from Fig. (6.7) that $v(k,y)$ attains its unique minimum on D at $k = -i\tau_s(y)$. We can now estimate (6.47) via the method of steepest descents. Accordingly, we first write, on the contour D close to $k = -i\tau_s(y)$,

$$k = k_d(\sigma) = \sigma + i\tau_d(\sigma) = \sigma - i\tau_s(y) + O\left(\sigma^2\right) \tag{6.49}$$

as $\sigma \to 0$, from which we approximate (6.47), via (6.38) and (6.49), as

$$J_+(y,t) \backsim \int_{-\delta(t)}^{\delta(t)} f(\sigma - i\tau_s(y)) \exp\left(-tv(\sigma - i\tau_s(y), y)\right) d\sigma \tag{6.50}$$

for $y > 1 + o(1)$ as $t \to \infty$ with $\delta(t) = O\left(t^{-\frac{2}{5}}\right)$ as $t \to \infty$. As $\sigma \to 0$ we have

$$f(\sigma - i\tau_s(y)) = -\frac{1}{\tau_s(y)^2}\left(1 - \exp\left(\beta\tau_s(y)\right)\right) - \frac{\beta}{\tau_s(y)} + O(\sigma) \tag{6.51}$$

and

$$v(\sigma - i\tau_s(y), y) = v(-i\tau_s(y), y) + \frac{1}{2}v_{kk}(-i\tau_s(y))\sigma^2 + O\left(\sigma^3\right) \tag{6.52}$$

where

$$v(-i\tau_s(y), y) = -(\tau_s(y)\tan(\tau_s(y)))^{\frac{1}{2}} + y\tau_s(y) \tag{6.53}$$

$$v_{kk}(-i\tau_s(y)) = -\frac{1}{4}(\tau_s(y)\tan(\tau_s(y)))^{-\frac{3}{2}}\left(\tan(\tau_s(y)) + \tau_s(y)\sec(\tau_s(y))^2\right)^2 \tag{6.54}$$

$$+ (\tau_s(y)\tan(\tau_s(y)))^{-\frac{1}{2}}\left(1 + \tau_s(y)\tan(\tau_s(y))\right)\sec(\tau_s(y))^2$$

with $\tau_s(y) \in (0, \frac{\pi}{2})$. Graphs of $v(-i\tau_s(y), y)$ and $v_{kk}(-i\tau_s(y))$ for $y > 1$ are shown in Figs. (6.9) and (6.10) respectively. We observe that $v(-i\tau_s(y), y) > 0$ and $v_{kk}(-i\tau_s(y)) > 0$ as expected. Thus, via (6.50), (6.51) and (6.52), we have

$$J_+(y,t) \backsim \exp\left(-tv(-i\tau_s(y), y)\right) \int_{-\delta(t)}^{\delta(t)} \left(-\frac{1}{\tau_s(y)^2}\left(1 - \exp\left(\beta\tau_s(y)\right)\right)\right.$$

$$\left. - \frac{\beta}{\tau_s(y)} + O(\sigma)\right) \times \exp\left(-\frac{1}{2}tv_{kk}(-i\tau_s(y))\sigma^2 + O\left(t\sigma^3\right)\right) d\sigma \tag{6.55}$$

for $y > 1 + o(1)$ as $t \to \infty$. We use the substitution

$$u^2 = \frac{1}{2}tv_{kk}(-i\tau_s(y))\sigma^2$$

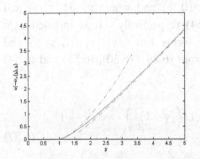

Fig. 6.9: The graph of $v(-i\tau_s(y),y)$ for $y > 1$ with the asymptotic approximations ($--$) from (6.58).

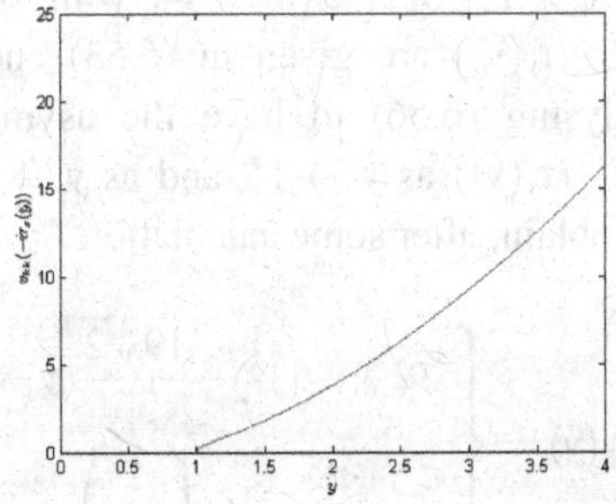

Fig. 6.10: The graph of $v_{kk}(-i\tau_s(y))$ for $y > 1$.

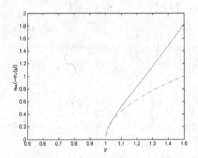

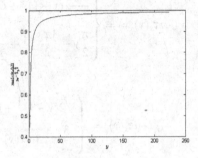

Fig. 6.11: The graph of $v_{kk}(-i\tau_s(y))$ for $y > 1$ with the asymptotic approximation ($--$) from (6.59) as $y \to 1^+$.

Fig. 6.12: Graph of the ratio between $v_{kk}(-i\tau_s(y))$ and the asymptotic form from (6.59) as $y \to \infty$.

in (6.55) to obtain, on setting $\delta(t) = t^{-\frac{2}{5}}$,

$$J_+(y,t) \backsim \exp\left(-tv(-i\tau_s(y),y)\right) \left(-\frac{1}{\tau_s(y)^2}\left(1 - \exp(\beta\tau_s(y))\right) - \frac{\beta}{\tau_s(y)}\right)$$

$$\times \sqrt{\frac{2}{tv_{kk}(-i\tau_s(y))}} \int_{-\infty}^{\infty} \exp\left(-u^2\right) du$$

for $y > 1 + o(1)$ as $t \to \infty$. Finally we have the approximation

$$J_+(y,t) \backsim \sqrt{\frac{2\pi}{tv_{kk}(-i\tau_s(y))}} \left(-\frac{\beta}{\tau_s(y)} - \frac{1}{\tau_s(y)^2}\left(1 - \exp(\beta\tau_s(y))\right)\right)$$

$$\times \exp\left(-tv(-i\tau_s(y),y)\right)$$

(6.56)

for $y > 1 + o(1)$ as $t \to \infty$, with $\tau_s(y) \in (0, \frac{\pi}{2})$ and where $v(-i\tau_s(y), y)$ and $v_{kk}(-i\tau_s(y))$ are given in (6.53) and (6.54) respectively. It is instructive in analysing (6.56) to have the asymptotic forms for $\tau_s(y)$, $v(-i\tau_s(y), y)$ and $v_{kk}(-i\tau_s(y))$ as $y \to 1^+$ and as $y \to \infty$. From (6.29), (6.30), (6.53) and (6.54) we obtain, after some calculation,

$$
\tau_s(y) = \begin{cases} \sqrt{2}(y-1)^{\frac{1}{2}} - \dfrac{19\sqrt{2}}{36}(y-1)^{\frac{3}{2}} + O\left((y-1)^{\frac{5}{2}}\right) & \text{as } y \to 1^+, \\[2ex] \dfrac{\pi}{2} - \dfrac{\pi^{\frac{1}{3}}}{2y^{\frac{2}{3}}} + O\left(\dfrac{1}{y^{\frac{4}{3}}}\right) & \text{as } y \to \infty. \end{cases} \tag{6.57}
$$

$$
v(-i\tau_s(y), y) = \begin{cases} \dfrac{2\sqrt{2}}{3}(y-1)^{\frac{3}{2}} + O\left((y-1)^{\frac{5}{2}}\right) & \text{as } y \to 1^+, \\[2ex] \dfrac{\pi}{2}y - \dfrac{3\pi^{\frac{1}{3}}}{2}y^{\frac{1}{3}} + \dfrac{1}{2\pi^{\frac{1}{3}}y^{\frac{1}{3}}} + O\left(\dfrac{1}{y}\right) & \text{as } y \to \infty. \end{cases} \tag{6.58}
$$

$$
v_{kk}(-i\tau_s(y)) = \begin{cases} \sqrt{2}(y-1)^{\frac{1}{2}} + O\left((y-1)^{\frac{3}{2}}\right) & \text{as } y \to 1^+, \\[2ex] \dfrac{3}{\pi^{\frac{1}{3}}}y^{\frac{5}{3}} + O(y) & \text{as } y \to \infty. \end{cases} \tag{6.59}
$$

The asymptotic forms (6.57) and (6.58) are included as dashed lines in Figs. (6.5) and (6.9) respectively, the asymptotic form (6.59) as $y \to 1^+$ is included as a dashed line in Fig. (6.11) and the graph of the ratio between $v_{kk}(-i\tau_s(y))$ and the asymptotic form (6.59) as $y \to \infty$ is shown in Fig. (6.12). It is again instructive to examine the form of (6.56) for $0 < y - 1 \ll 1$ and for $y \gg 1$. From (6.56), (6.57), (6.58) and (6.59) we obtain

$$
J_+(y,t) \sim \frac{2^{\frac{3}{2}}}{\sqrt{3}\pi^{\frac{1}{3}}t^{\frac{1}{2}}y^{\frac{5}{6}}}\left(\beta + \frac{2}{\pi}\left(\exp\left(\frac{\pi}{2}\beta\right) - 1\right)\right)\exp\left(-\frac{\pi}{2}t\left(y - \frac{3}{\pi^{\frac{1}{3}}}y^{\frac{1}{3}}\right)\right) \tag{6.60}
$$

with $y \gg 1$ as $t \to \infty$, and,

$$
J_+(y,t) \sim \frac{\sqrt{\pi}\beta^2}{2^{\frac{3}{4}}t^{\frac{1}{2}}(y-1)^{\frac{1}{4}}}\exp\left(-\frac{2\sqrt{2}}{3}t(y-1)^{\frac{3}{2}}\right) \tag{6.61}
$$

with $0 < (y-1) \ll 1$ as $t \to \infty$.

We now consider the approximation of $J_+(y,t)$ in Region II. Via (6.6) we have

$$J_+(y,t) = \int_{-\infty}^{\infty} f(k) \exp\left(itg(k,y)\right) dk, \qquad (6.62)$$

where

$$g(k,y) = \gamma(k) - yk \qquad (6.63)$$

for $y < 0$ with $t \in \mathbb{R}^+$. Setting $k = \sigma + i\tau$, with $\sigma, \tau \in \mathbb{R}$, we may write (6.63) as

$$g(k,y) = \gamma(k) - yk = u(k,y) + iv(k,y),$$

with $y < 0$, where $u(k,y)$ and $v(k,y)$ denote the real and imaginary parts of $g(k,y)$ respectively. Via (3.23), (3.24), (3.25) and (6.63), we may determine the qualitative behaviour of the level curves of $v(k,y)$, and a sketch of these level curves is given in Fig. (6.15), where the arrows point in the direction of increasing $u(k,y)$. We will deform the contour in (6.62) onto the contour E shown in Fig. (6.15), on which we write $v(k,y) = C(y)(> 0)$. On E, $k = k_e(\sigma) = \sigma + i\tau_e(\sigma)$, with $\tau_e(-\sigma) = \tau_e(\sigma)$, where $\tau_e(\sigma)$ is monotone increasing for $\sigma \geq 0$. Moreover,

$$\tau_e(\sigma) \to -\frac{C(y)}{y} \quad \text{as } |\sigma| \to \infty. \qquad (6.64)$$

for $y < 0$. Graphs of $C(y)$ and $y^{-1}C(y)$ are given in Figs. (6.13) and (6.14), and it is established numerically that $C(y) > 0$ for $y < 0$, $C(y)$ is monotone decreasing for $y < 0$, and

$$C(0) = 0.70324\ldots \qquad (6.65)$$

with,

$$C(y) \curvearrowleft -\frac{\pi}{2}y \qquad (6.66)$$

as $y \to -\infty$. We now consider the contour $D_M = [-M,M] \cup M_1 \cup M_2 \cup E_M$, where M_1 and M_2 are arcs on the circle $|k| = M$, and E_M is a finite section of E, as shown in Fig. (6.16). The points $k = p_1 = \sigma_{p_1} + i\tau_{p_1}$, and $k = p_2 = \sigma_{p_2} + i\tau_{p_2}$ are the intersection points of M_1 and M_2 with E_M respectively. The points $k = s_1 = \sigma_{s_1} + i\tau_{s_1}$ and $k = s_2 = \sigma_{s_2} + i\tau_{s_2}$ are the stationary points of $g(k,y)$ on the level curve E. We have, via (6.64), that

$$\sigma_{p_1} = M\left(1 - \frac{C(y)^2}{2y^2M^2} + O\left(\frac{C(y)^4}{y^4M^4}\right)\right) \qquad (6.67)$$

$$\tau_{p_1} = -\frac{C(y)}{y} + O\left(\frac{C(y)^2}{y^2M^2}\right) \qquad (6.68)$$

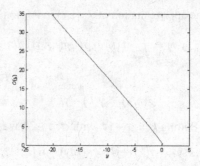

Fig. 6.13: The graph of $C(y)$ with $y < 0$.

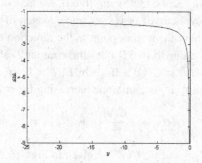

Fig. 6.14: The graph of $\frac{C(y)}{y}$ with $y < 0$.

as $M \to \infty$. With θ_M being the angle shown in Fig. (6.16), we also have from (6.67) and (6.68) that

$$\theta_M = -\frac{C(y)}{yM} + O\left(\frac{C(y)^2}{y^2 M^2}\right) \qquad (6.69)$$

as $M \to \infty$. Now, via Cauchy's Theorem, we have

$$\int_{D_M} f(k) \exp\left(itg(k,y)\right) dk = 0,$$

where

$$f(k) = \frac{1}{k^2}\left(1 - \exp\left(i\beta k\right)\right) + i\frac{\beta}{k} \qquad (6.70)$$

and

$$g(k,y) = \gamma(k) - yk = u(k,y) + iv(k,y), \qquad (6.71)$$

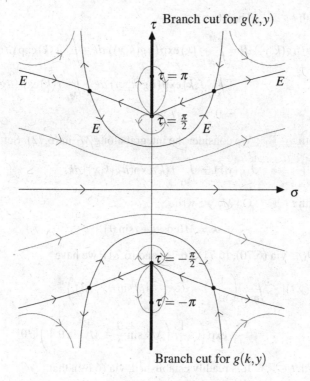

Fig. 6.15: Level curves for $v(k,y)$ in the cut k-plane, with $y < 0$.

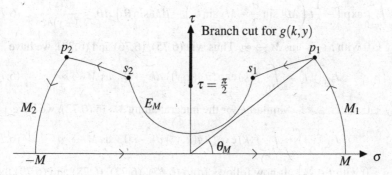

Fig. 6.16: Contour D_M.

for $y < 0$ with $t \in \mathbb{R}^+$. Therefore

$$\int_{E_M} f(k)\exp(itg(k,y))\,dk = \int_{-M}^{M} f(k)\exp(itg(k,y))\,dk + \int_{M_1} f(k)\exp(itg(k,y))\,dk$$

$$+ \int_{E_M} f(k)\exp(itg(k,y))\,dk + \int_{M_2} f(k)\exp(itg(k,y))\,dk$$

$$= 0 \tag{6.72}$$

for $y < 0$ with $t \in \mathbb{R}^+$. We consider the integral along M_1 in (6.72). Set

$$J_{M_1}(y,t) = \int_{M_1} f(k)\exp(itg(k,y))\,dk, \tag{6.73}$$

for $y < 0$ with $t \in \mathbb{R}^+$. On M_1 we write

$$k = M(\cos\theta + i\sin\theta) \tag{6.74}$$

for $\theta \in [0, \theta_M]$. Via (6.70), (6.71), (6.73) and (6.74), we have

$$|J_{M_1}(y,t)| \leq \int_{\theta=0}^{\theta_M} \left(\frac{1}{M}\left(1 + \exp(-\beta M \sin\theta) + \beta\right)\right)$$

$$\times \exp\left(-\frac{1}{2}t\left(M^{\frac{1}{2}}\sin\frac{\theta}{2} - My\sin\theta\right)\right)d\theta \tag{6.75}$$

for $y < 0$ with $t \in \mathbb{R}^+$. It is readily established, via (6.69), that

$$\int_{\theta=0}^{\theta_M} \exp\left(-\frac{1}{2}t\left(M^{\frac{1}{2}}\sin\frac{\theta}{2} - My\sin\theta\right)\right)d\theta \leq \frac{C(y)}{(-y)M} \tag{6.76}$$

and

$$\int_{\theta=0}^{\theta_M} \exp\left(-\frac{1}{2}t\left(M^{\frac{1}{2}}\sin\frac{\theta}{2} - My\sin\theta\right) - \beta M\sin\theta\right)d\theta \leq \frac{C(y)}{(-y)M} \tag{6.77}$$

for $y < 0$ with $t \in \mathbb{R}^+$ as $M \to \infty$. Thus, via (6.75), (6.76) and (6.77), we have

$$J_{M_1}(y,t) = \int_{M_1} f(k)\exp(itg(k,y))\,dk \to 0 \quad \text{as } M \to \infty \tag{6.78}$$

for $y < 0$ with $t \in \mathbb{R}^+$. Similarly, for the integral along M_2 in (6.72), we have

$$J_{M_2}(y,t) = \int_{M_2} f(k)\exp(itg(k,y))\,dk \to 0 \quad \text{as } M \to \infty \tag{6.79}$$

for $y < 0$ with $t \in \mathbb{R}^+$. It now follows from (6.62), (6.72), (6.78) and (6.79), that, on letting $M \to \infty$, we have

$$J_+(y,t) = -\int_E f(k)\exp(itg(k,y))\,dk \tag{6.80}$$

for $y < 0$ with $t \in \mathbb{R}^+$, where $f(k)$ and $g(k,y)$ are given by (6.70) and (6.71). Recall that on the contour E we have

$$g(k,y) = u(k,y) + iC(y) \tag{6.81}$$

so that, via (6.80) and (6.81), we have

$$J_+(y,t) = -\exp(-tC(y)) \int_E f(k) \exp(itu(k,y))\, dk \tag{6.82}$$

for $y < 0$ with $t \in \mathbb{R}^+$. In approximating $J_+(y,t)$, as $t \to \infty$, in (6.82), it is necessary to write the contour E as $E = E_1 \cup E_\Delta \cup E_2$, where $Re(k) \in (-\infty, -\Delta(t))$ on E_1, $Re(k) \in [-\Delta(t), \Delta(t)]$ on E_Δ and $Re(k) \in (\Delta(t), \infty)$ on E_2, with $\Delta(t) = 2 + t$, as shown in Fig. (6.17), so that

$$J_+(y,t) = -\exp(-tC(y)) \left(\int_{E_1} f(k)\exp(itu(k,y))\, dk + \int_{E_\Delta} f(k)\exp(itu(k,y))\, dk \right.$$

$$\left. + \int_{E_2} f(k)\exp(itu(k,y))\, dk \right) \tag{6.83}$$

for $y < 0$ with $t \in \mathbb{R}^+$. The points $k = s_1$ and $k = s_2$, shown in Fig. (6.17), are the stationary points of $u(k,y)$ on the contour E. Setting $k = \sigma(y) + i\tau(y)$ at the stationary point $k = s_1$, we may determine the position of the stationary point of $u(k,y)$ for $y < 0$, via (6.71). Graphs of $\sigma(y)$ and $\tau(y)$ at $k = s_1$, for $y < 0$, are given in Figs. (6.18) and (6.19) respectively. It is clear from Figs. (6.18) and (6.19) that,

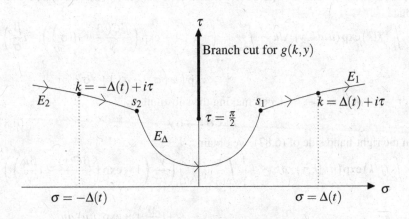

Fig. 6.17: Contour E.

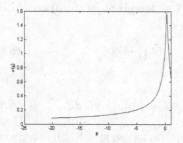 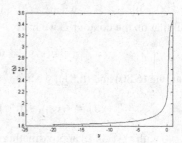

Fig. 6.18: The graph of $\sigma(y)$ at $k = s_1$ with $y < 0$.

Fig. 6.19: The graph of $\tau(y)$ at $k = s_1$ with $y < 0$.

for $y < 0$, since $\Delta(t) > 2$, the stationary points $k = s_1$ and $k = s_2$ are on the contour E_Δ. We now consider the approximation of the integral along E_Δ in (6.83). It follows from the Riemann-Lebesgue Lemma that

$$\int_{E_\Delta} f(k) \exp\left(itu(k,y)\right) dk \to 0 \tag{6.84}$$

for $y < 0$ as $t \to \infty$. We now consider the approximation on E_1 in (6.83). We may write, via (6.68), with t sufficiently large,

$$k \backsim \sigma - i\frac{C(y)}{y} \tag{6.85}$$

on E_1 with $y < 0$ and $\sigma \in (\Delta(t), \infty)$. Thus, via (3.25), (6.71) and (6.85), for t sufficiently large,

$$u(k,y) \backsim \sigma^{\frac{1}{2}} - \sigma y \tag{6.86}$$

on E_1 with $y < 0$ and $\sigma \in (\Delta(t), \infty)$. It then follows, via (6.70), (6.85) and (6.86), that

$$\int_{E_1} f(k) \exp\left(itu(k,y)\right) dk \backsim \int_{\sigma=\Delta(t)}^{\infty} \left(\frac{1}{\sigma^2}\left(1 - \exp\left(\frac{\beta C(y)}{y} + i\beta\sigma\right)\right) + i\frac{\beta}{\sigma}\right)$$

$$\times \exp\left(it\left(\sigma^{\frac{1}{2}} - \sigma y\right)\right) d\sigma$$

for $y < -o(1)$ as $t \to \infty$. Upon making the substitution

$$u = \sigma^{\frac{1}{2}} - \sigma y \tag{6.87}$$

in the right-hand side of (6.87), we obtain

$$\int_{E_1} f(k) \exp\left(itu(k,y)\right) dk \backsim -\frac{1}{y} \int_{\Delta(t)^{\frac{1}{2}} - y\Delta(t)}^{\infty} \left(\frac{y^2}{u^2}\left(1 - \exp\left(\frac{\beta C(y)}{y} - i\frac{\beta u}{y}\right)\right)\right.$$

$$\left. + i\frac{\beta y}{u}\right) \times \exp\left(itu\right) du$$

$$\tag{6.88}$$

for $y < -o(1)$ as $t \to \infty$. An integration by parts on the right-hand side of (6.88) establishes that

$$\int_{E_1} f(k) \exp(itu(k,y)) \, dk = O\left(\frac{1}{(-y)t^2}\right) \tag{6.89}$$

for $y < -o(1)$ as $t \to \infty$. Similarly, for the integral on E_2 in (6.83) it may be established that

$$\int_{E_2} f(k) \exp(itu(k,y)) \, dk = O\left(\frac{1}{(-y)t^2}\right) \tag{6.90}$$

for $y < -o(1)$ as $t \to \infty$. Therefore, we have, via (6.83), (6.84), (6.89) and (6.90), that

$$J_+(y,t) = o(\exp(-tC(y))) \tag{6.91}$$

for $y < -o(1)$ as $t \to \infty$. This completes the approximation of $J_+(y,t)$ in Region I, Region II and Region III.

Similarly, we may approximate $J_-(y,t)$ in (6.7). In this case there are three distinct asymptotic regions to consider, namely,

Region I′: $-1 + o(1) < y < -o(1)$ as $t \to \infty$,

Region II′: $y > o(1)$ as $t \to \infty$,

Region III′: $y < -1 - o(1)$ as $t \to \infty$.

In Region I′ we obtain,

$$J_-(y,t) \sim \sqrt{\frac{2\pi}{t\gamma''(-k_s(-y))}} \left(\frac{2\beta}{k_s(-y)} \sin\left(-\frac{\pi}{4} + tk_s(-y)(y + c(k_s(-y)))\right)\right.$$

$$+ \frac{2}{k_s(-y)^2}\left(\cos\left(-\frac{\pi}{4} + tk_s(-y)(y + c(k_s(-y)))\right)\right.$$

$$\left.\left. - \cos\left(-\frac{\pi}{4} - \beta k_s(-y) + tk_s(-y)(y + c(k_s(-y)))\right)\right)\right) \tag{6.92}$$

with $-1 + o(1) < y < -o(1)$ as $t \to \infty$, where

$$c(k) = \frac{\gamma(k)}{k},$$

for $k \geq 0$, and $k = \pm k_s(y)$ are the points where the phase becomes stationary in (6.7), with

$$k_s(-y) = \sqrt{2}(1+y)^{\frac{1}{2}} + O\left((1+y)^{\frac{3}{2}}\right) \tag{6.93}$$

as $y \to -1^+$, and

$$k_s(-y) = \frac{1}{4y^2} + O\left(\frac{1}{y^4}\exp\left(\frac{-1}{2y^2}\right)\right) \tag{6.94}$$

as $y \to 0^-$. Then, via (6.92), (6.93) and (6.94) we have,

$$J_-(y,t) \backsim -\frac{8\sqrt{\pi}(-y)^{\frac{1}{2}}}{t^{\frac{1}{2}}}\left(\beta\sin\left(\frac{\pi}{4}+\frac{t}{4y}\right) - 4y^2\left(\cos\left(\frac{\pi}{4}+\frac{t}{y}\right)\right.\right.$$
$$\left.\left. -\cos\left(\frac{\pi}{4}+\frac{\beta}{4y^2}+\frac{t}{4y}\right)\right)\right) \tag{6.95}$$

with $0 < (-y) \ll 1$ as $t \to \infty$, and

$$J_-(y,t) \backsim \frac{2^{\frac{1}{4}}\beta^2\sqrt{\pi}}{t^{\frac{1}{2}}(1+y)^{\frac{1}{4}}}\cos\left(\frac{\pi}{4} - \frac{2\sqrt{2}}{3}t(1+y)^{\frac{3}{2}}\right) \tag{6.96}$$

with $0 < (1+y) \ll 1$ as $t \to \infty$. In Region III$'$ we obtain,

$$J_-(y,t) \backsim \sqrt{\frac{2\pi}{tv_{kk}(-i\tau_s(-y))}}\left(\frac{\beta}{\tau_s(-y)} - \frac{1}{\tau_s(-y)^2}(1-\exp(-\beta\tau_s(-y)))\right)$$
$$\times \exp(tv(-i\tau_s(-y),y)) \tag{6.97}$$

for $y < -1 - o(1)$ as $t \to \infty$, where $k = \pm i\tau_s(-y)$ are the stationary points of $\gamma(k) + ky$ in (6.7), and $v(k) = Im(\gamma(k)+ky))$, with $v(-i\tau_s(-y),y) < 0$ and $v_{kk}(-i\tau_s(-y)) < 0$ for $y < -1$. Also,

$$J_-(y,t) \backsim \frac{2^{\frac{3}{2}}}{\sqrt{3}\pi^{\frac{1}{3}}t^{\frac{1}{2}}(-y)^{\frac{5}{6}}}\left(\beta + \frac{2}{\pi}\left(\exp\left(-\frac{\pi}{2}\beta\right) - 1\right)\right)$$
$$\times \exp\left(-\frac{\pi}{2}t\left((-y) - \frac{3}{\pi^{\frac{2}{3}}}(-y)^{\frac{1}{3}}\right)\right) \tag{6.98}$$

with $(-y) \gg 1$ as $t \to \infty$, and,

$$J_-(y,t) \backsim -\frac{\sqrt{\pi}\beta^2}{2^{\frac{3}{4}}t^{\frac{1}{2}}(-(y+1))^{\frac{1}{4}}}\exp\left(-\frac{2\sqrt{2}}{3}t(-(y+1))^{\frac{3}{2}}\right) \tag{6.99}$$

with $0 < (-(y+1)) \ll 1$ as $t \to \infty$. Finally, in Region II$'$ we obtain,

$$J_-(y,t) = o(\exp(tC(-y))) \tag{6.100}$$

for $y > o(1)$ as $t \to \infty$, where $C(-y) < 0$ for $y > 0$, with

$$C(-y) \backsim -\frac{\pi}{2}y$$

as $y \to \infty$, and

$$C(0) = -0.70324\ldots.$$

6.3   Outer Region Coordinate Expansions for $F(y,t)$ as $t \to \infty$

We now approximate $F(y,t)$, as given in (6.8), in Region I, Region II and Region III. We begin in Region I, where we write, via (6.8) and (6.12),

$$F(y,t) = \int_0^{\delta(t)} \frac{1}{k} \sin t \left(\gamma(k) - ky \right) dk + \int_{\delta(t)}^{\frac{1}{2} k_s(y)} \frac{1}{k} \sin t \left(\gamma(k) - ky \right) dk$$

$$+ \int_{\frac{1}{2} k_s(y)}^{\infty} \frac{1}{k} \sin t \left(\gamma(k) - ky \right) dk \tag{6.101}$$

for $0 < y < 1$ and $t \in \mathbb{R}^+$, where $\delta(t) = o(1)$ as $t \to \infty$ and $k = \pm k_s(y)$ are the stationary points of $\gamma(k) - ky$. We consider the first integral on the right-hand side of (6.101). Set

$$F_1(y,t) = \int_0^{\delta(t)} \frac{1}{k} \sin t \left(\gamma(k) - ky \right) dk \tag{6.102}$$

for $0 < y < 1$ and $t \in \mathbb{R}^+$, with $\delta(t) = o(1)$ as $t \to \infty$. Recall that

$$\gamma(k) = (k \tanh k)^{\frac{1}{2}} \tag{6.103}$$

for $-\infty < k < \infty$, so that, via (6.102) and (6.103), and on taking $\delta(t) = o \left(t^{-\frac{1}{3}} \right)$ as $t \to \infty$, we have

$$F_1(y,t) = \int_0^{\delta(t)} \frac{1}{k} \sin \left(t(1-y)k \right) dk - \frac{t}{6} \int_0^{\delta(t)} k^2 \cos \left(t(1-y)k \right) dk + O \left(t \delta(t)^5 \right) \tag{6.104}$$

for $0 < y < 1$ as $t \to \infty$. Upon making the substitution

$$w = t(1-y)k \tag{6.105}$$

in (6.104), and taking $t\delta(t) \to \infty$ as $t \to \infty$, we obtain

$$F_1(y,t) = \int_0^{\infty} \frac{\sin w}{w} dw - \int_{t(1-y)\delta(t)}^{\infty} \frac{\sin w}{w} dw$$

$$- \frac{1}{6t^2(1-y)^3} \int_0^{t(1-y)\delta(t)} w^2 \cos w \, dw + O \left(t\delta(t)^5 \right) \tag{6.106}$$

for $0 < y < 1$ as $t \to \infty$. Integration by parts and using the result

$$\int_0^{\infty} \frac{\sin w}{w} dw = \frac{\pi}{2} \tag{6.107}$$

we finally have

$$F_1(y,t) = \frac{\pi}{2} - \frac{\cos \left(t\delta(t)(1-y) \right)}{t\delta(t)(1-y)} + O \left(\frac{\delta(t)^2}{(1-y)} + \frac{1}{t^2\delta(t)^2(1-y)^2} + t\delta(t)^5 \right) \tag{6.108}$$

for $0 < y < 1$ with $\delta(t) = o\left(t^{-\frac{1}{3}}\right)$ and $t\delta(t) \to \infty$ as $t \to \infty$. We now consider the second integral on the right-hand side of (6.101). Set

$$F_2(y,t) = \int_{\delta(t)}^{\frac{1}{2}k_s(y)} \frac{1}{k} \sin t \left(\gamma(k) - ky\right) dk \qquad (6.109)$$

for $0 < y < 1$ and $t \in \mathbb{R}^+$, where $k = k_s(y)$ is a stationary point of $\gamma(k) - ky$. It is convenient to write (6.109) as

$$F_2(y,t) = \frac{1}{t} \int_{\delta(t)}^{\frac{1}{2}k_s(y)} \frac{t\left(\gamma'(k) - y\right)}{k\left(\gamma'(k) - y\right)} \sin t \left(\gamma(k) - ky\right) dk \qquad (6.110)$$

for $0 < y < 1$ and $t \in \mathbb{R}^+$. After an integration by parts in (6.110) we obtain

$$F_2(y,t) = \frac{\cos\left(t\delta(t)(1-y)\right)}{t\delta(t)(1-y)} - \frac{1}{t} \int_{\delta(t)}^{\frac{1}{2}k_s(y)} \left(\frac{1}{k^2\left(\gamma'(k) - y\right)} + \frac{\gamma''(k)}{k\left(\gamma'(k) - y\right)^2} \right)$$

$$\times \cos t \left(\gamma(k) - ky\right) dk + O\left(\frac{\delta(t)^2}{(1-y)} + \frac{1}{t} \right) \qquad (6.111)$$

for $0 < y < 1$ as $t \to \infty$. By successive integration by parts in (6.111), we observe that

$$F_2(y,t) = \frac{\cos\left(t\delta(t)(1-y)\right)}{t\delta(t)(1-y)} + O\left(\frac{1}{t} + \frac{\delta(t)^2}{(1-y)} + \frac{1}{t^2\delta(t)^2(1-y)^2} \right) \qquad (6.112)$$

for $0 < y < 1$ with $\delta(t) = o\left(t^{-\frac{1}{3}}\right)$ and $t\delta(t) \to \infty$ as $t \to \infty$. We now consider the third integral on the right-hand side of (6.101). Set

$$F_3(y,t) = \int_{\frac{1}{2}k_s(y)}^{\infty} \frac{1}{k} \sin t \left(\gamma(k) - ky\right) dk \qquad (6.113)$$

for $0 < y < 1$ and $t \in \mathbb{R}^+$. It is convenient to write (6.113) as

$$F_3(y,t) = \frac{1}{2i} \int_{\frac{1}{2}k_s(y)}^{\infty} \frac{1}{k} \exp\left(it\left(\gamma(k) - ky\right)\right) dk - \frac{1}{2i} \int_{\frac{1}{2}k_s(y)}^{\infty} \frac{1}{k} \exp\left(-it\left(\gamma(k) - ky\right)\right) dk \qquad (6.114)$$

for $0 < y < 1$ and $t \in \mathbb{R}^+$, and we consider each integral in turn. We consider the first integral on the right-hand side of (6.114). Set

$$F_+(y,t) = \int_{\frac{1}{2}k_s(y)}^{\infty} \frac{1}{k} \exp\left(it\left(\gamma(k) - ky\right)\right) \qquad (6.115)$$

for $0 < y < 1$ and $t \in \mathbb{R}^+$. The phase is stationary at $k = k_s(y)$, so we may approximate $F_+(y,t)$ as

$$F_+(y,t) \sim \frac{1}{k_s(y)} \exp\left(it\left(\gamma(k_s(y)) - yk_s(y)\right)\right) \int_{-\varepsilon(t)}^{\varepsilon(t)} \exp\left(i\frac{1}{2}t\left(\gamma''(k_s(y))u^2\right)\right) du \qquad (6.116)$$

for $o(1) < y < 1 - o(1)$ as $t \to \infty$, with $\varepsilon(t) = o\left(t^{-\frac{1}{3}}\right)$ as $t \to \infty$, and we have used the substitution $u = k - k_s(y)$. As $\gamma''(k) < 0$ for $k > 0$, and $\gamma''(-k) = -\gamma''(k)$, we use the substitution

$$u^2 = \frac{1}{2} t \gamma''(-k_s(y)) w^2$$

in (6.116) to obtain, on taking $\varepsilon(t) = t^{-\frac{2}{3}}$,

$$F_+(y,t) \backsim \frac{1}{k_s(y)} \sqrt{\frac{2}{t\gamma''(-k_s(y))}} \exp\left(it\left(\gamma(k_s(y)) - yk_s(y)\right)\right) \int_{-\infty}^{\infty} \exp\left(-iw^2\right) dw$$

for $o(1) < y < 1 - o(1)$ as $t \to \infty$. Using the result in (6.21) we have

$$F_+(y,t) \backsim \frac{1}{k_s(y)} \sqrt{\frac{2\pi}{t\gamma''(-k_s(y))}} \exp\left(i\left(t\left(\gamma(k_s(y)) - yk_s(y)\right) - \frac{\pi}{4}\right)\right) \quad (6.117)$$

for $o(1) < y < 1 - o(1)$ as $t \to \infty$. Similarly, for the second integral in (6.114) we have

$$F_-(y,t) = \int_{\frac{1}{2}k_s(y)}^{\infty} \frac{1}{k} \exp\left(it\left(\gamma(k) - ky\right)\right)$$

$$\backsim \frac{1}{k_s(y)} \sqrt{\frac{2\pi}{t\gamma''(-k_s(y))}} \exp\left(-i\left(t\left(\gamma(k_s(y)) - yk_s(y)\right) - \frac{\pi}{4}\right)\right)$$

$$(6.118)$$

for $o(1) < y < 1 - o(1)$ as $t \to \infty$. Hence, via (6.114), (6.115), (6.117) and (6.118), we obtain

$$F_3(y,t) \backsim -\frac{1}{k_s(y)} \sqrt{\frac{2\pi}{t\gamma''(-k_s(y))}} \sin\left(\frac{\pi}{4} + tk_s(y)\left(y - c(k_s(y))\right)\right) \quad (6.119)$$

for $o(1) < y < 1 - o(1)$ as $t \to \infty$, where

$$c(k) = \frac{\gamma(k)}{k} \quad (6.120)$$

for $k \geq 0$. Finally, via (6.101), (6.102), (6.108), (6.109), (6.112), (6.113) and (6.119), we have

$$F(y,t) \backsim \frac{\pi}{2} - \frac{1}{k_s(y)} \sqrt{\frac{2\pi}{t\gamma''(-k_s(y))}} \sin\left(\frac{\pi}{4} + tk_s(y)\left(y - c(k_s(y))\right)\right) \quad (6.121)$$

for $o(1) < y < 1 - o(1)$ as $t \to \infty$. It is now instructive to consider the form of (6.121) when $o(1) < y \ll 1$ and $o(1) < 1 - y \ll 1$. We have, via (6.13) and (6.15), that

$$k_s(y) = \sqrt{2}(1-y)^{\frac{1}{2}} + o\left((1-y)^{\frac{3}{2}}\right) \quad (6.122)$$

as $y \to 1^-$, and

$$k_s(y) = \frac{1}{4y^2} + O\left(\frac{1}{y^4} \exp\left(\frac{-1}{2y^2}\right)\right) \tag{6.123}$$

as $y \to 0^+$. Thus, via (3.24), (3.25), (6.120), (6.121), (6.122) and (6.123), it is readily established that

$$F(y,t) \sim \frac{\pi}{2} - \frac{4\pi^{\frac{1}{2}}y^{\frac{1}{2}}}{t^{\frac{1}{2}}} \sin\left(\frac{\pi}{4} - \frac{t}{4y}\right) \tag{6.124}$$

when $o(1) < y \ll 1$ as $t \to \infty$, and

$$F(y,t) \sim \frac{\pi}{2} - \frac{\pi^{\frac{1}{2}}}{2^{\frac{1}{4}}t^{\frac{1}{2}}(1-y)^{\frac{3}{4}}} \sin\left(\frac{\pi}{4} - \frac{2\sqrt{2}}{3}t(1-y)^{\frac{3}{2}}\right) \tag{6.125}$$

when $o(1) < (1-y) \ll 1$ as $t \to \infty$.

We now consider the approximation of $F(y,t)$ in Region III. Via (6.8) we may write

$$F(y,t) = \int_0^{\delta(t)} \frac{1}{k} \sin t(\gamma(k) - ky) \, dk + \int_{\delta(t)}^{\frac{1}{2}\tau_s(y)} \frac{1}{k} \sin t(\gamma(k) - ky) \, dk$$

$$+ \int_{\frac{1}{2}\tau_s(y)}^{\infty} \frac{1}{k} \sin t(\gamma(k) - ky) \, dk \tag{6.126}$$

for $y > 1$ with $t \in \mathbb{R}^+$, where $k = \pm i\tau_s(y)$ are the stationary points of $\gamma(k) - ky$ on $k = i\tau$, as given in (6.30), and $\delta(t) = o(1)$ as $t \to \infty$. Following the methodology approach in (6.102)–(6.112) it is established that

$$\int_0^{\delta(t)} \frac{1}{k} \sin t(\gamma(k) - ky) \, dk + \int_{\delta(t)}^{\frac{1}{2}\tau_s(y)} \frac{1}{k} \sin t(\gamma(k) - ky) \, dk = -\frac{\pi}{2} + O\left(\frac{1}{t}\right) \tag{6.127}$$

for $y > 1 + o(1)$ with $\delta(t) = o\left(t^{-\frac{1}{3}}\right)$ as $t \to \infty$. We now consider the third integral on the right-hand side of (6.126). Set

$$F_4(y,t) = \int_{\frac{1}{2}\tau_s(y)}^{\infty} \frac{1}{k} \sin t(\gamma(k) - ky) \, dk \tag{6.128}$$

$$= \frac{1}{2i}\left(F_{+\tau_s}(y,t) - F_{-\tau_s}(y,t)\right)$$

for $y > 1$ with $t \in \mathbb{R}^+$, where

$$F_{+\tau_s}(y,t) = \int_{\frac{1}{2}\tau_s(y)}^{\infty} \frac{1}{k} \exp\left(it(\gamma(k) - ky)\right) dk \tag{6.129}$$

and

$$F_{-\tau_s}(y,t) = \int_{\frac{1}{2}\tau_s(y)}^{\infty} \frac{1}{k} \exp\left(-it(\gamma(k) - ky)\right) dk \qquad (6.130)$$

We will first consider the approximation of $F_{+\tau_s}(y,t)$. Setting $k = \sigma + i\tau$, with $\sigma \geq 0$ and $\tau \in \mathbb{R}$, we may write

$$\gamma(k) - ky = u(k,y) + iv(k,y) \qquad (6.131)$$

with $y > 1$. Using (6.31) and Fig. (6.7) we have the qualitative behaviour of the level curves of $u(k,y)$, for $\sigma > 0$, as shown in Fig. (6.20), where the arrows point in the direction of increasing $v(k,y)$. We observe that the level curve D_- is the path of steepest descent for $\gamma(k) - ky$ in (6.129). Thus we will deform the contour in (6.129) onto the steepest descent contour D_-.

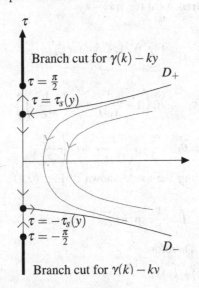

Fig. 6.20: Level curves for $u(k,y)$ in (6.131) in the cut k-plane, with $y > 1$.

On D_-, $k = k_{d_-}(\sigma) = \sigma + i\tau_{d_-}(\sigma)$, with $\tau_{d_-}(0) = -\tau_s(y)$. We now consider the contour $C_{L_-} = \left[\frac{1}{2}\tau_s(y), L\right] \cup L_1 \cup D_{L_-} \cup D_2 \cup D_1$, where L_1 and D_2 are arcs on the circle $|k| = L$ and $|k| = \frac{1}{2}\tau_s(y)$ respectively, and D_{L_-} is a finite section of D_-, as shown in Fig. (6.21). The point $k = k_1 = \sigma_1 + i\tau_1$ is the intersection point of L_1 with D_{L_-}, and we recall, via (6.33), (6.34) and (6.35), that

$$\sigma_1 = \frac{L^{\frac{1}{2}}}{\sqrt{2}y} \left(1 - \frac{1}{8y^2L} + O\left(\frac{1}{y^4L^2}\right)\right),$$

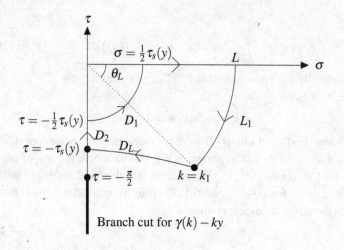

Fig. 6.21: Contour C_{L_-}.

$$\tau_1 = -L\left(1 - \frac{1}{4y^2 L} + O\left(\frac{1}{y^4 L^2}\right)\right),$$

$$\theta_L = \frac{\pi}{2} - \frac{1}{\sqrt{2}yL^{\frac{1}{2}}} + O\left(\frac{1}{y^3 L^{\frac{3}{2}}}\right) \tag{6.132}$$

as $L \to \infty$, with θ_L being the angle shown in Fig. (6.21). Now, via Cauchy's Theorem, we have

$$\int_{C_{L_-}} \frac{1}{k} \exp\left(it\left(\gamma(k) - ky\right)\right) dk = 0$$

for $y > 1$ with $t \in \mathbb{R}^+$. Therefore

$$\int_{C_{L_-}} \frac{1}{k} \exp\left(it\left(\gamma(k) - ky\right)\right) dk$$

$$= \int_{\frac{1}{2}\tau_s(y)}^{L} \frac{1}{k} \exp\left(it\left(\gamma(k) - ky\right)\right) dk + \int_{L_1} \frac{1}{k} \exp\left(it\left(\gamma(k) - ky\right)\right) dk$$

$$+ \int_{D_{L_-}} \frac{1}{k} \exp\left(it\left(\gamma(k) - ky\right)\right) dk + \int_{L_{\tau_s}} \frac{1}{k} \exp\left(it\left(\gamma(k) - ky\right)\right) dk \tag{6.133}$$

$$+ \int_{L_{\frac{1}{2}\tau_s}} \frac{1}{k} \exp\left(it\left(\gamma(k) - ky\right)\right) dk$$

$$= 0$$

for $y > 1$ with $t \in \mathbb{R}^+$. We now consider the integral along L_1 in (6.133). Set

$$F_{L_1}(y,t) = \int_{L_1} \frac{1}{k} \exp\left(it\left(\gamma(k) - ky\right)\right) dk \qquad (6.134)$$

for $y > 1$ with $t \in \mathbb{R}^+$. On L_1 we write

$$k = L(\cos\theta + i\sin\theta) \qquad (6.135)$$

for $\theta \in [-\theta_L, 0]$. Via (3.25), (6.132), (6.134) and (6.135) we have

$$|F_{L_1}(y,t)| \leq \int_{\theta=-\frac{\pi}{2}}^{0} \exp\left(t\left(yL\sin\theta - 2L^{\frac{1}{2}}\sin\frac{\theta}{2}\right)\right) d\theta. \qquad (6.136)$$

Thus, via (6.43) and (6.136), we have

$$F_{L_1}(y,t) = \int_{L_1} \frac{1}{k} \exp\left(it\left(\gamma(k) - ky\right)\right) dk \to 0 \qquad (6.137)$$

as $L \to \infty$, for $y > 1$ with $t \in \mathbb{R}^+$. It now follows from (6.129), (6.133) and (6.137) that, on letting $L \to \infty$, we have

$$F_{+\tau_s}(y,t) = -\int_{D_-} \frac{1}{k} \exp\left(it\left(\gamma(k) - ky\right)\right) dk - \int_{D_1} \frac{1}{k} \exp\left(it\left(\gamma(k) - ky\right)\right) dk$$

$$- \int_{D_2} \frac{1}{k} \exp\left(it\left(\gamma(k) - ky\right)\right) dk$$

$$(6.138)$$

for $y > 1$ with $t \in \mathbb{R}^+$. We now consider the integral along D_1 in (6.138). Set

$$F_{D_1}(y,t) = \int_{D_1} \frac{1}{k} \exp\left(it\left(\gamma(k) - ky\right)\right) dk \qquad (6.139)$$

for $y > 1$ with $t \in \mathbb{R}^+$. On D_1 we write

$$k = \frac{1}{2}\tau_s(y)\exp(i\theta) \qquad (6.140)$$

for $\theta \in \left[-\frac{\pi}{2}, 0\right]$. Via (6.139) and (6.140) we have

$$F_{D_1}(y,t) = i\int_{\theta=-\frac{\pi}{2}}^{0} \exp\left(it\left(\gamma\left(\frac{1}{2}\tau_s(y)\exp(i\theta)\right) - \frac{1}{2}\tau_s(y)y\exp(i\theta)\right)\right) d\theta$$

$$(6.141)$$

for $y > 1$ with $t \in \mathbb{R}^+$. On integrating by parts in (6.141) we observe that

$$F_{D_1}(y,t) = O\left(\frac{1}{t}\right) \qquad (6.142)$$

for $y > 1 + o(1)$ as $t \to \infty$. We next consider the integral along D_2 in (6.138). Set

$$F_{D_2}(y,t) = \int_{D_2} \frac{1}{k} \exp\left(it\left(\gamma(k) - ky\right)\right) dk \qquad (6.143)$$

for $y > 1$ with $t \in \mathbb{R}^+$. On D_2 we write

$$k = i\tau \tag{6.144}$$

for $\tau \in \left[-\tau_s(y), -\frac{1}{2}\tau_s(y)\right]$. Via (6.143) and (6.144) we have

$$F_{D_2}(y,t) = -\int_{\tau=\frac{1}{2}\tau_s(y)}^{\tau_s(y)} \frac{1}{\tau} \exp\left(-it\left(\gamma(i\tau) - i\tau y\right)\right) d\tau \tag{6.145}$$

for $y > 1$ with $t \in \mathbb{R}^+$. We now consider the integral along D_- in (6.138). Set

$$F_{D_-}(y,t) = \int_{D_-} \frac{1}{k} \exp\left(it\left(\gamma(k) - ky\right)\right) dk \tag{6.146}$$

for $y > 1$ with $t \in \mathbb{R}^+$. Via (6.131) and recalling that $u(k,y) = 0$ on D_-, we have $\gamma(k) - ky = iv(k,y)$ on D_-. The level curve D_- is the path of steepest descent for $\gamma(k) - ky$ in (6.146), and we observe from Fig. (6.21) that $v(k,y)$ attains its unique minimum on D_- at $k = -i\tau_s(y)$. We can now estimate (6.146) via the method of steepest descents. We write on the contour D_-, close to $k = -i\tau_s(y)$,

$$k = \sigma - i\tau_s(y) + O\left(\sigma^2\right) \tag{6.147}$$

as $\sigma \to 0$, from which we approximate (6.146), via (6.147), as

$$F_{D_-}(y,t) \backsim \frac{\exp\left(-tv(-i\tau_s(y),y)\right)}{i\tau_s(y)} \int_0^{\varepsilon(t)} \exp\left(-\frac{1}{2}tv_{kk}\left(-i\tau_s(y)\right)\sigma^2\right) d\sigma \tag{6.148}$$

for $y > 1 + o(1)$ as $t \to \infty$ with $\varepsilon(t) = O\left(t^{-\frac{2}{5}}\right)$ as $t \to \infty$, where $v(-i\tau_s(y),y)$ and $v_{kk}(-i\tau_s(y))$ are given in (6.53) and (6.54) and in Figs. (6.9) and (6.10) respectively. Upon making the substitution

$$u^2 = \frac{1}{2}tv_{kk}(-i\tau_s(y))\sigma^2$$

in (6.148) and setting $\varepsilon(t) = t^{-\frac{2}{5}}$ we obtain

$$F_{D_-}(y,t) \backsim \frac{1}{i\tau_s(y)} \sqrt{\frac{2}{tv_{kk}(-i\tau_s(y))}} \exp\left(-tv(-i\tau_s(y),y)\right) \int_0^\infty \exp(-u^2) du \tag{6.149}$$

for $y > 1 + o(1)$ as $t \to \infty$. Using the result

$$\int_0^\infty \exp\left(-u^2\right) du = \frac{\sqrt{\pi}}{2}$$

in (6.149) we have

$$F_{D_-}(y,t) \backsim \frac{1}{i\tau_s(y)} \sqrt{\frac{\pi}{2tv_{kk}(-i\tau_s(y))}} \exp\left(-tv(-i\tau_s(y),y)\right) \tag{6.150}$$

for $y > 1 + o(1)$ as $t \to \infty$. Thus, via (6.138), (6.142), (6.145) and (6.150), we have

$$F_{+\tau_s}(y,t) \backsim \frac{i}{\tau_s(y)} \sqrt{\frac{\pi}{2tv_{kk}(-i\tau_s(y))}} \exp(-tv(-i\tau_s(y),y))$$

$$+ \int_{\tau=\frac{1}{2}\tau_s(y)}^{\tau_s(y)} \frac{1}{\tau} \exp\left(-it\left(\gamma(i\tau) - i\tau y\right)\right) d\tau + O\left(\frac{1}{t}\right) \tag{6.151}$$

for $y > 1 + o(1)$ as $t \to \infty$. Similarly, we may approximate $F_{-\tau_s}(y,t)$ given in (6.130) as

$$F_{-\tau_s}(y,t) \backsim - \frac{i}{\tau_s(y)} \sqrt{\frac{\pi}{2tv_{kk}(-i\tau_s(y))}} \exp(-tv(-i\tau_s(y),y))$$

$$+ \int_{\tau=\frac{1}{2}\tau_s(y)}^{\tau_s(y)} \frac{1}{\tau} \exp\left(-it\left(\gamma(i\tau) - i\tau y\right)\right) d\tau + O\left(\frac{1}{t}\right) \tag{6.152}$$

for $y > 1 + o(1)$ as $t \to \infty$. Thus via (6.128), (6.151) and (6.152) we have

$$\int_{\frac{1}{2}\tau_s(y)}^{\infty} \frac{1}{k} \sin t(\gamma(k) - ky) \, dk \backsim \frac{1}{\tau_s(y)} \sqrt{\frac{\pi}{2tv_{kk}(-i\tau_s(y))}} \exp(-tv(-i\tau_s(y),y)) \tag{6.153}$$

for $y > 1 + o(1)$ as $t \to \infty$. Finally, we have, via (6.126), (6.127) and (6.153),

$$F(y,t) \backsim -\frac{\pi}{2} + \frac{1}{\tau_s(y)} \sqrt{\frac{\pi}{2tv_{kk}(-i\tau_s(y))}} \exp(-tv(-i\tau_s(y),y)) \tag{6.154}$$

for $y > 1 + o(1)$ as $t \to \infty$. It is instructive to examine the form of (6.154) for $0 < y - 1 \ll 1$ and for $y \gg 1$. From (6.57), (6.58) and (6.59) we obtain

$$F(y,t) \backsim -\frac{\pi}{2} + \frac{\sqrt{2}}{\sqrt{3}\pi^{\frac{1}{3}}t^{\frac{1}{2}}y^{\frac{5}{6}}} \exp\left(-\frac{\pi}{2}t\left(y - \frac{3}{\pi^{\frac{2}{3}}}y^{\frac{1}{3}} + \frac{1}{\pi^{\frac{4}{3}}y^{\frac{1}{3}}}\right)\right) . \tag{6.155}$$

with $y \gg 1$ as $t \to \infty$, and,

$$F(y,t) \backsim -\frac{\pi}{2} + \frac{\sqrt{\pi}}{2^{\frac{5}{4}}t^{\frac{1}{2}}(y-1)^{\frac{3}{4}}} \exp\left(-\frac{2\sqrt{2}}{3}t(y-1)^{\frac{3}{2}}\right) \tag{6.156}$$

for $0 < (y - 1) \ll 1$ as $t \to \infty$.

We now consider the approximation of $F(y,t)$ in Region II. Via (6.8) we may write

$$F(y,t) = \int_0^{\delta(t)} \frac{1}{k} \sin t(\gamma(k) - ky) \, dk + \int_{\delta(t)}^{\frac{1}{2}} \frac{1}{k} \sin t(\gamma(k) - ky) \, dk$$

$$+ \int_{\frac{1}{2}}^{\infty} \frac{1}{k} \sin t(\gamma(k) - ky) \, dk \tag{6.157}$$

for $y < 0$ with $t \in \mathbb{R}^+$, where $\delta(t) = o(1)$ as $t \to \infty$. Following the approach in (6.102)–(6.112) it is established that

$$\int_0^{\delta(t)} \frac{1}{k} \sin t(\gamma(k) - ky)\, dk + \int_{\delta(t)}^{\frac{1}{2}} \frac{1}{k} \sin t(\gamma(k) - ky)\, dk = \frac{\pi}{2} + O\left(\frac{1}{t}\right) \quad (6.158)$$

for $y < -o(1)$ with $\delta(t) = o\left(t^{-\frac{1}{3}}\right)$ as $t \to \infty$. We now consider the third integral on the right-hand side of (6.158). Set

$$F_5(y,t) = \int_{\frac{1}{2}}^{\infty} \frac{1}{k} \sin t(\gamma(k) - ky)\, dk$$

$$= \frac{1}{2i} \left(F_{5+}(y,t) - F_{5-}(y,t)\right) \quad (6.159)$$

for $y < 0$ with $t \in \mathbb{R}^+$, where

$$F_{5+}(y,t) = \int_{\frac{1}{2}}^{\infty} \frac{1}{k} \exp\left(i\left(t(\gamma(k) - ky)\right)\right) dk \quad (6.160)$$

and

$$F_{5-}(y,t) = \int_{\frac{1}{2}}^{\infty} \frac{1}{k} \exp\left(-i\left(t(\gamma(k) - ky)\right)\right) dk \quad (6.161)$$

We will first consider the approximation of $F_{5+}(y,t)$. Setting $k = \sigma + i\tau$, with $\sigma \geq 0$ and $\tau \in \mathbb{R}$, we may write

$$\gamma(k) - ky = u(k,y) + iv(k,y) \quad (6.162)$$

with $y < 0$. Via (6.162) and Fig. (6.15) we have the qualitative behaviour of the level curves of $v(k,y)$ for $\sigma > 0$, as shown in Fig. (6.22), where the arrows point in the direction of increasing $u(k,y)$. We will deform the contour in (6.160) onto the contour E_+ shown in Fig. (6.22), on which we write $v(k,y) = C(y)(> 0)$. On E_+ we have $k = \sigma + i\tau_{e_+}(\sigma)$, where $\tau_{e_+}(\sigma)$ is monotone increasing for $\sigma \geq 0$. We recall, via (6.64), (6.65) and (6.66), that

$$\tau_{e_+}(\sigma) \to -\frac{C(y)}{y} \quad \text{as } \sigma \to \infty$$

and

$$C(0) = 0.70324\ldots$$

with,

$$C(y) \backsim -\frac{\pi}{2} y$$

as $y \to -\infty$. Graphs of $C(y)$ and $y^{-1}C(y)$ are given in Figs. (6.13) and (6.14) respectively. We now consider the contour $D_{M+} = \left[\frac{1}{2}, M\right] \cup M_1 \cup E_{M+} \cup E_1 \cup E_2$,

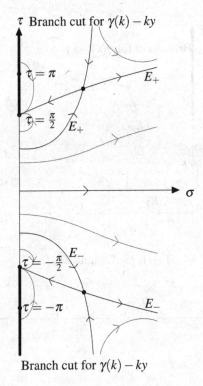

Fig. 6.22: Level curves for $v(k,y)$, in (6.162), in the cut k-plane, with $y < 0$.

where M_1 is an arc on the circle $|k| = M$, E_2 is an arc on the circle $|k| = \frac{1}{2}$ and E_{M+} is a finite section of E_+ as shown in Fig. (6.23). The point $k = i\tau_{C(y)}$ is the intersection point of E_{M+} with E_1, where $\tau_{C(y)}$ is monotone decreasing for $y < 0$ and it is established numerically that $\frac{1}{2} < \tau_{C(y)} < \frac{\pi}{2}$ for $y < 0$, and a graph of $\tau_{C(y)}$ is given in Fig. (6.24). The point $k = p = \sigma_p + i\tau_p$ is the intersection point of M_1 with E_{M+}, and we recall, via (6.67), (6.68) and (6.69), that

$$\sigma_p = M\left(1 - \frac{C(y)^2}{2y^2M^2} + O\left(\frac{C(y)^4}{y^4M^4}\right)\right),$$

$$\tau_p = -\frac{C(y)}{y} + O\left(\frac{C(y)^2}{y^2M^2}\right),$$

$$\theta_M = -\frac{C(y)}{yM} + O\left(\frac{C(y)^2}{y^2M^2}\right)$$

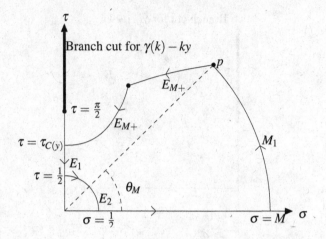

Fig. 6.23: Contour D_{M+}.

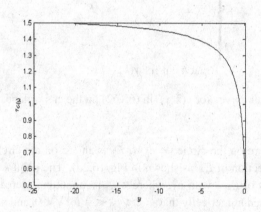

Fig. 6.24: The graph of $\tau_{C(y)}$ for $y < 0$.

as $M \to \infty$, with θ_M being the angle shown in Fig. (6.23). Now, via Cauchy's Theorem, we have

$$\int_{D_{M+}} \frac{1}{k} \exp\left(it\left(\gamma(k) - ky\right)\right) dk = 0$$

for $y < 0$ with $t \in \mathbb{R}^+$. Therefore,

$$\int_{D_{M+}} \frac{1}{k} \exp\left(it\left(\gamma(k) - ky\right)\right) dk$$

$$= \int_{\frac{1}{2}}^{M} \frac{1}{k} \exp\left(it\left(\gamma(k) - ky\right)\right) dk + \int_{M_1} \frac{1}{k} \exp\left(it\left(\gamma(k) - ky\right)\right) dk$$

$$+ \int_{E_{M+}} \frac{1}{k} \exp\left(it\left(\gamma(k) - ky\right)\right) dk + \int_{E_1} \frac{1}{k} \exp\left(it\left(\gamma(k) - ky\right)\right) dk \qquad (6.163)$$

$$+ \int_{E_2} \frac{1}{k} \exp\left(it\left(\gamma(k) - ky\right)\right) dk$$

$$= 0$$

for $y < 0$ with $t \in \mathbb{R}^+$. We now consider the integral along M_1 in (6.163). Set

$$F_{M_1}(y,t) = \int_{M_1} \frac{1}{k} \exp\left(it\left(\gamma(k) - ky\right)\right) dk \qquad (6.164)$$

for $y < 0$ with $t \in \mathbb{R}^+$. On M_1 we write

$$k = M(\cos\theta + i\sin\theta) \qquad (6.165)$$

for $\theta \in [0, \theta_M]$. Via (6.76), (6.164) and (6.165), we have

$$F_{M_1}(y,t) \leq \int_{\theta=0}^{\theta_M} \exp\left(-\frac{1}{2}t\left(M^{\frac{1}{2}}\sin\frac{\theta}{2} - My\sin\theta\right)\right) d\theta \qquad (6.166)$$

$$\leq \frac{C(y)}{(-y)M}$$

for $y < 0$ with $t \in \mathbb{R}^+$ as $M \to \infty$. It now follows from (6.163) and (6.166) that, on letting $M \to \infty$, we have

$$F_{5+}(y,t) = -\int_{E_+} \frac{1}{k} \exp\left(it\left(\gamma(k) - ky\right)\right) dk - \int_{E_1} \frac{1}{k} \exp\left(it\left(\gamma(k) - ky\right)\right) dk$$

$$- \int_{E_2} \frac{1}{k} \exp\left(it\left(\gamma(k) - ky\right)\right) dk$$

$$(6.167)$$

for $y < 0$ with $t \in \mathbb{R}^+$. We now consider the integral along E_1 in (6.167). Set

$$F_{E_1}(y,t) = \int_{E_1} \frac{1}{k} \exp\left(it\left(\gamma(k) - ky\right)\right) dk \qquad (6.168)$$

for $y < 0$ with $t \in \mathbb{R}^+$. On E_1 we write

$$k = i\tau \qquad (6.169)$$

for $\tau \in \left[\frac{1}{2}, \tau_{C(y)}\right]$. Via (6.168) and (6.169) we have

$$F_{E_1}(y,t) = -\int_{\frac{1}{2}}^{\tau_{C(y)}} \frac{1}{\tau} \exp\left(it\left(\gamma(i\tau) - i\tau y\right)\right) d\tau \qquad (6.170)$$

for $y < 0$ with $t \in \mathbb{R}^+$. We next consider the integral along E_2 in (6.167). Set

$$F_{E_2}(y,t) = \int_{E_2} \frac{1}{k} \exp\left(it\left(\gamma(k) - ky\right)\right) dk \qquad (6.171)$$

for $y < 0$ with $t \in \mathbb{R}^+$. On E_2 we write

$$k = \frac{1}{2} \exp\left(i\theta\right) \qquad (6.172)$$

for $\theta \in \left[0, \frac{\pi}{2}\right]$. Via (6.171) and (6.172), we have

$$F_{E_2}(y,t) = -i \int_{\theta=0}^{\frac{\pi}{2}} \exp\left(it\left(\gamma\left(\frac{1}{2}\exp\left(i\theta\right)\right) - \frac{1}{2}y\exp\left(i\theta\right)\right)\right) d\theta \qquad (6.173)$$

for $y < 0$ with $t \in \mathbb{R}^+$. On integrating by parts in (6.173) we observe that

$$F_{E_2}(y,t) = O\left(\frac{1}{t}\right) \qquad (6.174)$$

for $y < -o(1)$ as $t \to \infty$. We now consider the integral along E_+ in (6.167). Set

$$F_{E_+}(y,t) = \int_{E_+} \frac{1}{k} \exp\left(it\left(\gamma(k) - ky\right)\right) dk \qquad (6.175)$$

for $y < 0$ with $t \in \mathbb{R}^+$. Recall that on the contour E_+ we have

$$\gamma(k) - ky = u(k,y) + iC(y) \qquad (6.176)$$

so that, via (6.175) and (6.176), we have

$$F_{E_+}(y,t) = \exp(-tC(y)) \int_{E_+} \frac{1}{k} \exp(itu(k,y)) dk \qquad (6.177)$$

for $y < 0$ with $t \in \mathbb{R}^+$. In approximating $F_{E_+}(y,t)$, as $t \to \infty$, in (6.177), it is necessary to write the contour E_+ as $E_+ = E_\Delta \cup E_3$, where $Re(k) \in [0, \Delta(t)]$ on E_Δ and $Re(k) \in (\Delta(t), \infty)$ on E_3, with $\Delta(t) = t + 2$, as shown in Fig. (6.25), so that

$$F_{E_+}(y,t) = \exp(-tC(y)) \left(\int_{E_\Delta} \frac{1}{k} \exp(itu(k,y)) dk + \int_{E_3} \frac{1}{k} \exp(itu(k,y)) dk\right) \qquad (6.178)$$

for $y < 0$ with $t \in \mathbb{R}^+$. The point $k = s$ shown in Fig. (6.25) is the stationary point of $u(k,y)$ on the contour E_+. Setting $k = \sigma(y) + i\tau(y)$ at the stationary point $k = s$, we may determine the position of the stationary point of $u(k,y)$, for $y < 0$ via (6.176). Graphs of $\sigma(y)$ and $\tau(y)$ at $k = s_1$ for $y < 0$, are given in Figs. (6.18) and (6.19) respectively. It is clear from Figs. (6.18) and (6.19) that, for $y < 0$, since

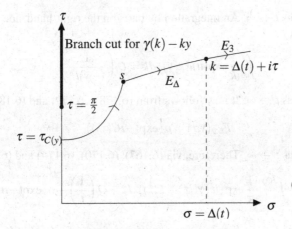

Fig. 6.25: Contour E_+.

$\Delta(t) > 2$, the stationary point $k = s_1$ is on the contour E_Δ. We now consider the integral along E_Δ in (6.178). It follows from the Riemann-Lebesgue Lemma that

$$\int_{E_\Delta} \frac{1}{k} \exp(itu(k,y))\, dk \to 0 \tag{6.179}$$

for $y < -o(1)$ as $t \to \infty$. We now consider the integral on E_3 in (6.178). We may write, via (6.68), with t sufficiently large,

$$k \sim \sigma - i\frac{C(y)}{y} \tag{6.180}$$

on E_3 with $y < 0$ and $\sigma \in (\Delta(t), \infty)$. Thus, via (3.25), (6.176) and (6.180), for t sufficiently large,

$$u(k,y) \sim \sigma^{\frac{1}{2}} - \sigma y \tag{6.181}$$

on E_3 with $y < 0$ and $\sigma \in (\Delta(t), \infty)$. It then follows, via (6.180) and (6.181), that

$$\int_{E_3} \frac{1}{k} \exp(itu(k,y))\, dk \sim \int_{\sigma=\Delta(t)}^{\infty} \frac{1}{\sigma} \exp\left(it\left(\sigma^{\frac{1}{2}} - y\sigma\right)\right) d\sigma \tag{6.182}$$

for $y < -o(1)$ as $t \to \infty$. Upon making the substitution

$$u = \sigma^{\frac{1}{2}} - \sigma y$$

in the right-hand side of (6.182), we obtain

$$\int_{E_3} \frac{1}{k} \exp(itu(k,y))\, dk \sim \int_{\sigma=\Delta(t)^{\frac{1}{2}} - y\Delta(t)}^{\infty} \frac{1}{u} \exp(itu)\, du \tag{6.183}$$

for $y < -o(1)$ as $t \to \infty$. An integration by parts on the right-hand side of (6.183) establishes that

$$\int_{E_3} \frac{1}{k} \exp(itu(k,y))\, dk = O\left(\frac{1}{(-y)t^2}\right) \qquad (6.184)$$

for $y < -o(1)$ as $t \to \infty$. It now follows from (6.178), (6.179) and (6.184), that

$$F_{E_+}(y,t) = o\left(\exp(-tC(y))\right) \qquad (6.185)$$

for $y < -o(1)$ as $t \to \infty$. Therefore, via (6.167), (6.170), (6.174) and (6.185) that

$$F_{5+}(y,t) = \int_{\frac{1}{2}}^{\tau_{C(y)}} \frac{1}{\tau} \exp\left(it\left(\gamma(i\tau) - i\tau y\right)\right) d\tau + O\left(\frac{1}{t}\right) + o\left(\exp(-tC(y))\right) \qquad (6.186)$$

for $y < -o(1)$ as $t \to \infty$. Similarly, we may approximate $F_{5-}(y,t)$ in (6.161) as

$$F_{5-}(y,t) = \int_{\frac{1}{2}}^{\tau_{C(y)}} \frac{1}{\tau} \exp\left(it\left(\gamma(i\tau) - i\tau y\right)\right) d\tau + O\left(\frac{1}{t}\right) + o\left(\exp(-tC(y))\right) \qquad (6.187)$$

for $y < -o(1)$ as $t \to \infty$. Thus, via (6.159), (6.186) and (6.187), we have

$$F_5(y,t) = O\left(\frac{1}{t}\right) + o\left(\exp(-tC(y))\right) \qquad (6.188)$$

for $y < -o(1)$ as $t \to \infty$. Finally, via (6.157), (6.158), (6.159) and (6.188), we have

$$F(y,t) = \frac{\pi}{2} + o\left(\exp(-tC(y))\right) \qquad (6.189)$$

for $y < -o(1)$ as $t \to \infty$.

6.4   Outer Region Coordinate Expansions for $\bar{\eta}(y,t)$ as $t \to \infty$

We can now construct the approximation for $\bar{\eta}(y,t)$ in each region. We have, via (6.5),

$$\bar{\eta}(y,t) = \frac{1}{2} + \frac{1}{4\pi\beta}\left(J_+(y,t) + J_-(y,t)\right) + \frac{1}{2\pi}\left(F(y,t) - F(-y,t)\right) \qquad (6.190)$$

for $(y,t) \in \mathbb{R} \times \mathbb{R}^+$. Thus, via (6.24), (6.25), (6.26), (6.56), (6.60), (6.61), (6.91), (6.92), (6.95), (6.96), (6.97), (6.98), (6.99), (6.100), (6.121), (6.124), (6.125), (6.154), (6.155), (6.156), (6.189) and (6.190) we have the following approximation for $\bar{\eta}(y,t)$ in the following regions,

Region I$^+$: $o(1) < y \left(= \frac{x}{t}\right) < 1 - o(1)$ as $t \to \infty$. We have,

$$\bar{\eta}(y,t) \backsim \frac{1}{2} + \frac{1}{\beta\sqrt{2\pi t\gamma''(-k_s(y))}k_s(y)^2} \left(\cos\left(\frac{\pi}{4} + tk_s(y)\,(y - c(k_s(y)))\right) \right.$$

$$\left. - \cos\left(\frac{\pi}{4} - \beta k_s(y) + tk_s(y)\,(y - c(k_s(y)))\right) \right) \right) \qquad \text{as } t \to \infty,$$

$$(6.191)$$

with

$$c(k) = \frac{\gamma(k)}{k} \qquad (6.192)$$

and $k = k_s(y)$ is the positive stationary point of $\gamma(k) - yk$ when $o(1) < y < 1 - o(1)$. In particular,

$$\bar{\eta}(y,t) \backsim \frac{1}{2} + \frac{8y^{\frac{5}{2}}}{\beta\pi^{\frac{1}{2}}t^{\frac{1}{2}}} \left(\cos\left(\frac{\pi}{4} - \frac{t}{4y}\right) - \cos\left(\frac{\pi}{4} - \frac{\beta}{4y^2} - \frac{t}{4y}\right) \right). \qquad (6.193)$$

when $0 < y \ll 1$ as $t \to \infty$, and

$$\bar{\eta}(y,t) \backsim \frac{1}{2} - \frac{1}{2^{\frac{5}{4}}\pi^{\frac{1}{2}}t^{\frac{1}{2}}} \left(\frac{1}{(1-y)^{\frac{3}{4}}} \sin\left(\frac{\pi}{4} - \frac{2\sqrt{2}}{3}t(1-y)^{\frac{3}{2}}\right) \right.$$

$$(6.194)$$

$$\left. - \frac{\beta}{\sqrt{2}(1-y)^{\frac{1}{4}}} \cos\left(\frac{\pi}{4} - \frac{2\sqrt{2}}{3}t(1-y)^{\frac{3}{2}}\right) \right)$$

when $0 < (1-y) \ll 1$ as $t \to \infty$.

Region II$^+$: $y \left(= \frac{x}{t}\right) > 1 + o(1)$ as $t \to \infty$. We have,

$$\bar{\eta}(y,t) \backsim \frac{1}{2\beta\sqrt{2\pi t v_{kk}(-i\tau_s(y))}\tau_s(y)^2} \left(\exp\left(\beta\tau_s(y)\right) - 1\right) \exp\left(-tv\left(-i\tau_s(y),y\right)\right)$$

$$(6.195)$$

as $t \to \infty$, where $k = \pm i\tau_s(y)$ (with $\tau_s(y) > 0$) are the stationary points of $\gamma(k) - yk$ when $y > 1 + o(1)$. In particular,

$$\bar{\eta}(y,t) \backsim \frac{1}{2^{\frac{9}{4}}\pi^{\frac{1}{2}}t^{\frac{1}{2}}} \left(\frac{1}{(y-1)^{\frac{3}{4}}} + \frac{\beta}{\sqrt{2}(y-1)^{\frac{1}{4}}} \right) \exp\left(-\frac{2\sqrt{2}}{3}t(y-1)^{\frac{3}{2}}\right) \quad (6.196)$$

when $0 < (y-1) \ll 1$ as $t \to \infty$, and

$$\bar{\eta}(y,t) \backsim \frac{1}{2\pi\beta} \left(\exp\left(\frac{\pi}{2}\beta\right) - 1\right) \left(\frac{4}{\sqrt{6}\pi^{\frac{4}{3}}t^{\frac{1}{2}}y^{\frac{5}{6}}} \right) \exp\left(-\frac{\pi}{2}t\left(y - \frac{3}{\pi^{\frac{2}{3}}}y^{\frac{1}{3}}\right)\right)$$

$$(6.197)$$

when $y \gg 1$ as $t \to \infty$, and it is readily verified that (5.92) and (6.197) asymptotically match according to the asymptotic matching principal of Van Dyke [Van Dyke (1975)].

Region I$^-$: $-1 + o(1) < y \left(= \frac{x}{t} \right) < -o(1)$ as $t \to \infty$. We have,

$$\bar{\eta}(y,t) \backsim \frac{1}{2} + \frac{1}{\beta \sqrt{2\pi t \gamma''(-k_s(-y))} k_s(-y)^2} \left(\cos\left(-\frac{\pi}{4} + tk_s(-y)(y + c(k_s(-y)))\right) \right.$$

$$\left. - \cos\left(-\frac{\pi}{4} - \beta k_s(-y) + tk_s(-y)(y + c(k_s(-y)))\right) \right) \right) \quad \text{as } t \to \infty,$$

(6.198)

with

$$c(k) = \frac{\gamma(k)}{k} \tag{6.199}$$

where $k = k_s(-y)$ is the positive stationary point of $\gamma(k) + yk$ when $-1 + o(1) < y < -o(1)$. In particular,

$$\bar{\eta}(y,t) \backsim \frac{1}{2} + \frac{8(-y)^{\frac{5}{2}}}{\beta \pi^{\frac{1}{2}} t^{\frac{1}{2}}} \left(\cos\left(\frac{\pi}{4} + \frac{t}{4y}\right) - \cos\left(\frac{\pi}{4} + \frac{\beta}{4y^2} + \frac{t}{4y}\right) \right) \tag{6.200}$$

when $0 < (-y) \ll 1$ as $t \to \infty$, and

$$\bar{\eta}(y,t) \backsim \frac{1}{2} + \frac{1}{2^{\frac{5}{4}} \pi^{\frac{1}{2}} t^{\frac{1}{2}}} \left(\frac{1}{(1+y)^{\frac{3}{4}}} \sin\left(\frac{\pi}{4} - \frac{2\sqrt{2}}{3} t(1+y)^{\frac{3}{2}}\right) \right.$$

$$\left. + \frac{\beta}{\sqrt{2}(1+y)^{\frac{1}{4}}} \cos\left(\frac{\pi}{4} - \frac{2\sqrt{2}}{3} t(1+y)^{\frac{3}{2}}\right) \right)$$

(6.201)

when $0 < (1+y) \ll 1$ as $t \to \infty$.

Region II$^-$: $y \left(= \frac{x}{t} \right) < -1 - o(1)$ as $t \to \infty$. We have,

$$\bar{\eta}(y,t) \backsim 1 - \frac{1}{2\beta \sqrt{2\pi t v_{kk}(-i\tau_s(-y))} \tau_s(-y)^2} (1 - \exp(\beta \tau_s(-y)))$$

$$\times \exp(-tv(-i\tau_s(-y), y)),$$

(6.202)

as $t \to \infty$, where $k = \pm i\tau_s(-y)$ (with $\tau_s(-y) > 0$) are the stationary points of $\gamma(k) + yk$ when $y < -1 - o(1)$. In particular,

$$\bar{\eta}(y,t) \backsim 1 - \frac{1}{2^{\frac{9}{4}} \pi^{\frac{1}{2}} t^{\frac{1}{2}}} \left(\frac{1}{(-(y+1))^{\frac{3}{4}}} - \frac{\beta}{\sqrt{2}(-(y+1))^{\frac{1}{4}}} \right)$$

$$\times \exp\left(-\frac{2\sqrt{2}}{3} t(-(y+1))^{\frac{3}{2}}\right)$$

(6.203)

when $0 < (-(y+1)) \ll 1$ as $t \to \infty$, and

$$\bar{\eta}(y,t) \sim 1 - \frac{1}{2\pi\beta}\left(1 - \exp\left(-\frac{\pi}{2}\beta\right)\right)\left(\frac{4}{\sqrt{6}\pi^{\frac{4}{3}}t^{\frac{1}{2}}(-y)^{\frac{5}{6}}}\right)\exp\left(\frac{\pi}{2}t\left(y - \frac{3}{\pi^{\frac{2}{3}}}y^{\frac{1}{3}}\right)\right)$$
(6.204)

when $(-y) \gg 1$ as $t \to \infty$, and it is readily verified that (5.93) and (6.204) asymptotically match according to the asymptotic matching principal of Van Dyke [Van Dyke (1975)].

Graphs of $\bar{\eta}(y,t)$ in each region (as determined from (6.191)–(6.204)) are given in Figs. (6.26)–(6.38), where we have set $t = 10$ and $\beta = 1$.

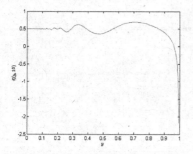

Fig. 6.26: The graph of $\bar{\eta}(y,10)$ in Region I$^+$, (6.191).

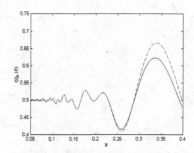

Fig. 6.27: The graph of $\bar{\eta}(y,10)$ in Region I$^+$, (6.191), with asymptotic approximation $(--)$, (6.193).

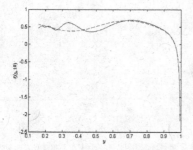

Fig. 6.28: The graph of $\bar{\eta}(y,10)$ in Region I$^+$, (6.191), with asymptotic approximation $(--)$, (6.194).

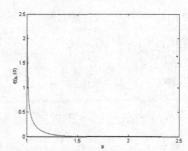

Fig. 6.29: The graph of $\bar{\eta}(y,10)$ in Region II$^+$, (6.195).

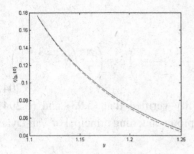

Fig. 6.30: The graph of $\bar{\eta}(y, 10)$ in Region II$^+$, (6.195), with asymptotic approximation (−−), (6.196).

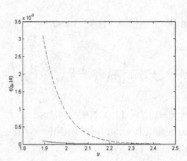

Fig. 6.31: The graph of $\bar{\eta}(y, 10)$ in Region II$^+$, (6.195), with asymptotic approximation (−−), (6.197).

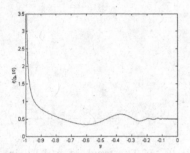

Fig. 6.32: The graph of $\bar{\eta}(y, 10)$ in Region I$^-$, (6.198).

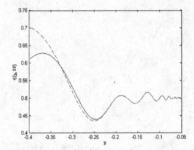

Fig. 6.33: The graph of $\bar{\eta}(y, 10)$ in Region I$^-$, (6.198), with asymptotic approximation (−−), (6.200).

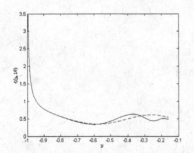

Fig. 6.34: The graph of $\bar{\eta}(y, 10)$ in Region I$^-$, (6.198), with asymptotic approximation (−−), (6.201).

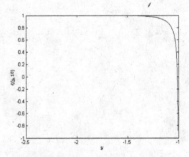

Fig. 6.35: The graph of $\bar{\eta}(y, 10)$ in Region II$^-$, (6.202).

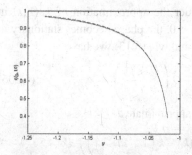

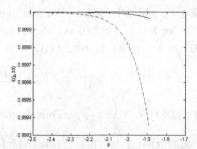

Fig. 6.36: The graph of $\bar{\eta}(y,10)$ in Region II$^-$, (6.202), with asymptotic approximation (— —), (6.203).

Fig. 6.37: The graph of $\bar{\eta}(y,10)$ in Region II$^-$, (6.202), with asymptotic approximation (— —), (6.204).

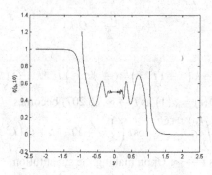

Fig. 6.38: The graph of $\bar{\eta}(y,10)$.

6.5 Inner Region Coordinate Expansions for $J_+(y,t)$ and $J_-(y,t)$ as $t \to \infty$

We now consider $J_+(y,t)$ in (6.6) with $y = 1 + O\left(t^{-\frac{2}{3}}\right)$ as $t \to \infty$, which we refer to as

Region $\hat{\mathbf{I}}^+$: $y = 1 \pm O\left(t^{-\frac{2}{3}}\right)$ as $t \to \infty$.

We write $y = 1 + Yt^{-\frac{2}{3}}$ with $Y = O(1)$ as $t \to \infty$. We now have, via (6.6),

$$J_+(Y,t) = \int_{-\infty}^{\infty} f(k) \exp\left(it\left(\gamma(k) - k\left(1 + Yt^{-\frac{2}{3}}\right)\right)\right) dk \qquad (6.205)$$

which we wish to estimate for $Y = O(1)$ as $t \to \infty$, where

$$f(k) = \frac{1}{k^2}\left(1 - \exp\left(i\beta k\right)\right) + i\frac{\beta}{k}$$

for $k \in \mathbb{C}$, which is entire. We first consider the approximation of $J_+(Y,t)$ in (6.205) for $Y = O(1)(<0)$ as $t \to \infty$. For $Y < 0$, the phase becomes stationary in (6.205) at $k = \pm k_s(Y,t)$, where $k_s(Y,t) > 0$, and, via (6.13), we have

$$k_s(Y,t) = \frac{\sqrt{2}(-Y)^{\frac{1}{2}}}{t^{\frac{1}{3}}} + O\left(\frac{(-Y)^{\frac{3}{2}}}{t}\right) \tag{6.206}$$

for $Y = O(1)(<0)$ as $t \to \infty$. Thus we may approximate $J_+(Y,t)$ as

$$J_+(Y,t) \backsim \int_{-k_s(Y,t)-\delta(t)}^{0} f(k) \exp\left(it\left(\gamma(k) - k\left(1 + Yt^{-\frac{2}{3}}\right)\right)\right) dk$$

$$+ \int_{0^-}^{k_s(Y,t)+\delta(t)} f(k) \exp\left(it\left(\gamma(k) - k\left(1 + Yt^{-\frac{2}{3}}\right)\right)\right) dk \tag{6.207}$$

for $Y = O(1)(<0)$ as $t \to \infty$, where $\delta(t) = o(1)$ as $t \to \infty$. For $|k| \ll 1$ we have

$$f(k) = \frac{\beta^2}{2} + O(k) \tag{6.208}$$

and

$$\gamma(k) - k\left(1 + Yt^{-\frac{2}{3}}\right) = -\frac{1}{6}k^3 - Yt^{-\frac{2}{3}}k + O\left(k^5\right)$$

so that, as $k_s(Y,t) + \delta(t) = o(1)$ as $t \to \infty$, (6.207) becomes

$$J_+(Y,t) \backsim \beta^2 \int_0^{k_s(Y,t)+\delta(t)} \cos t \left(\frac{1}{6}k^3 + Yt^{-\frac{2}{3}}k + O\left(k^5\right)\right) dk \tag{6.209}$$

for $Y = O(1)(<0)$ as $t \to \infty$. Upon making the substitution

$$v = \frac{k}{k_s(Y,t)} - 1$$

in (6.209) we obtain, via (6.206), and on taking $\delta(t) = t^{-\frac{1}{4}}$,

$$J_+(Y,t) \backsim \frac{\sqrt{2}\beta^2(-Y)^{\frac{1}{2}}}{t^{\frac{1}{3}}} \int_{v=-1}^{\infty} \cos\left(\left(2^{\frac{1}{3}}(-Y)\right)^{\frac{3}{2}}\left(\frac{1}{3}v^3 + v^2 - \frac{2}{3}\right)\right) dv \tag{6.210}$$

for $Y = O(1)(<0)$ as $t \to \infty$. The integral in (6.210) is now in the form of an Airy function, [Oliver *et al.* (2010)] (Chap. 9, p. 196). Hence we may write

$$J_+(Y,t) \backsim \frac{2^{\frac{1}{3}}\pi\beta^2}{t^{\frac{1}{3}}} Ai\left(2^{\frac{1}{3}}Y\right) \tag{6.211}$$

for $Y = O(1)(<0)$ as $t \to \infty$. We now consider the estimation of $J_+(Y,t)$ for $Y = O(1)(>0)$ as $t \to \infty$. For $Y > 0$ the phase becomes stationary in (6.205) at $k = \pm i\tau_s(Y,t)$, where $\tau_s(Y,t) > 0$, and, via (6.57), we have

$$\tau_s(Y,t) = \frac{\sqrt{2}Y^{\frac{1}{2}}}{t^{\frac{1}{3}}} + O\left(\frac{Y^{\frac{3}{2}}}{t}\right)$$

for $Y = O(1)(> 0)$ as $t \to \infty$. Thus we will deform the contour in (6.205) onto the steepest descent contour D, as shown in Fig. (6.7). Following the approach in (6.36)–(6.50), we may approximate $J_+(Y,t)$ as

$$J_+(Y,t) \sim \int_{\sigma=-\delta(t)}^{\delta(t)} f(\sigma - i\tau_s(Y,t))$$

$$\times \exp\left(it\left(\gamma(\sigma - i\tau_s(Y,t)) - (\sigma - i\tau_s(Y,t))\left(1 + Yt^{-\frac{2}{3}}\right)\right)\right) d\sigma \tag{6.212}$$

for $Y > 0$ as $t \to \infty$, where we have let $k = \sigma - i\tau_s(Y,t)$, with $\sigma \in (-\delta(t), \delta(t))$, on the contour D close to $k = -i\tau_s(Y,t)$, and $\delta(t) = o(1)$ as $t \to \infty$. For $|\sigma| \ll 1$ we have

$$\gamma(\sigma - i\tau_s(Y,t)) - (\sigma - i\tau_s(Y,t))\left(1 + Yt^{-\frac{2}{3}}\right)$$
$$= -\frac{1}{6}\sigma^3 + \frac{1}{\sqrt{2}}iY^{\frac{1}{2}}t^{-\frac{1}{3}}\sigma^2 + \frac{2\sqrt{2}}{3}iY^{\frac{3}{2}}t^{-1} + O\left(\sigma^5\right) \tag{6.213}$$

so that, via (6.208) and (6.213), (6.212) becomes

$$J_+(Y,t) \sim \frac{\beta^2}{2}\exp\left(-\frac{2\sqrt{2}}{3}Y^{\frac{3}{2}}\right)$$

$$\times \int_{\sigma=-\delta(t)}^{\delta(t)} \exp\left(-\frac{1}{6}it\sigma^3 - \frac{1}{\sqrt{2}}Y^{\frac{1}{2}}t^{\frac{2}{3}}\sigma^2 + O\left(t\sigma^5\right)\right) d\sigma \tag{6.214}$$

for $Y > 0$ as $t \to \infty$. Making the substitution

$$u = \frac{t^{\frac{1}{3}}}{2^{\frac{1}{3}}}\sigma$$

in (6.214) we have, on taking $\delta(t) = t^{-\frac{1}{4}}$,

$$J_+(Y,t) \sim \frac{2^{\frac{1}{3}}\beta^2}{t^{\frac{1}{3}}}\exp\left(-\frac{2\sqrt{2}}{3}Y^{\frac{3}{2}}\right)\int_{u=0}^{\infty}\exp\left(-\left(2^{\frac{1}{2}}Y\right)^{\frac{1}{3}}u^2\right)\cos\left(\frac{1}{3}u^3\right) du \tag{6.215}$$

for $Y = O(1)(> 0)$ as $t \to \infty$. The integral in (6.215) is now in the form of an Airy function, [Oliver *et al.* (2010)] (Chap. 9, p. 196). Hence we may write

$$J_+(Y,t) \sim \frac{2^{\frac{1}{3}}\pi\beta^2}{t^{\frac{1}{3}}}Ai\left(2^{\frac{1}{3}}Y\right) \tag{6.216}$$

for $Y = O(1)(> 0)$ as $t \to \infty$. Thus, via (6.211) and (6.216), we have

$$J_+(Y,t) \sim \frac{2^{\frac{1}{3}}\pi\beta^2}{t^{\frac{1}{3}}}Ai\left(2^{\frac{1}{3}}Y\right) \tag{6.217}$$

for $Y = O(1)$ as $t \to \infty$. In particular, we have, by [Oliver *et al.* (2010)] (Chap. 9, p. 198),

$$J_+(Y,t) \backsim \frac{2^{\frac{1}{4}}\sqrt{\pi}\beta^2}{t^{\frac{1}{3}}(-Y)^{\frac{1}{4}}} \cos\left(\frac{\pi}{4} - \frac{2\sqrt{2}}{3}(-Y)^{\frac{3}{2}}\right) \tag{6.218}$$

for $(-Y) \gg 1$ as $t \to \infty$, and

$$J_+(Y,t) \backsim \frac{\sqrt{\pi}\beta^2}{2^{\frac{3}{4}}t^{\frac{1}{3}}Y^{\frac{1}{4}}} \exp\left(-\frac{2\sqrt{2}}{3}Y^{\frac{3}{2}}\right) \tag{6.219}$$

for $Y \gg 1$ as $t \to \infty$. It is readily established that (6.218) and (6.219) asymptotically match accordingly with (6.26) (as $Y \to -\infty$) and (6.61) (as $Y \to +\infty$) respectively.

We next consider $J_-(y,t)$ in (6.7) with $y = -1 + O\left(t^{-\frac{2}{3}}\right)$ as $t \to \infty$, which we refer to as

Region $\hat{I}^-$: $y = -1 \pm O\left(t^{-\frac{2}{3}}\right)$ as $t \to \infty$.

In a similar manner to above, we obtain

$$J_-\left(\widehat{Y},t\right) \backsim \frac{2^{\frac{1}{3}}\pi\beta^2}{t^{\frac{1}{3}}}Ai\left(-2^{\frac{1}{3}}\widehat{Y}\right) \tag{6.220}$$

for $\widehat{Y} = O(1)$ as $t \to \infty$, with $y = -1 + \widehat{Y}t^{-\frac{2}{3}}$. In particular, we have,

$$J_-\left(\widehat{Y},t\right) \backsim \frac{2^{\frac{1}{4}}\sqrt{\pi}\beta^2}{t^{\frac{1}{3}}\widehat{Y}^{\frac{1}{4}}} \cos\left(\frac{\pi}{4} - \frac{2\sqrt{2}}{3}\widehat{Y}^{\frac{3}{2}}\right) \tag{6.221}$$

for $\widehat{Y} \gg 1$ as $t \to \infty$, and

$$J_-\left(\widehat{Y},t\right) \backsim \frac{\sqrt{\pi}\beta^2}{2^{\frac{3}{4}}t^{\frac{1}{3}}(-\widehat{Y})^{\frac{1}{4}}} \exp\left(-\frac{2\sqrt{2}}{3}(-\widehat{Y})^{\frac{3}{2}}\right) \tag{6.222}$$

for $\left(-\widehat{Y}\right) \gg 1$ as $t \to \infty$, and it is readily established that (6.221) $\left(\text{as } \widehat{Y} \to -\infty\right)$ and (6.222) $\left(\text{as } \widehat{Y} \to +\infty\right)$ asymptotically match accordingly with (6.96) and (6.99) respectively.

6.6   Inner Region Coordinate Expansions for $F(y,t)$ as $t \to \infty$

We now consider $F(y,t)$ in (6.8) in Region $\hat{I}^+$, where we write $y = 1 + Yt^{-\frac{2}{3}}$. We now have, via (6.8),

$$F(Y,t) = \int_0^\infty \frac{1}{k}\sin t\left(\gamma(k) - k\left(1 + Yt^{-\frac{2}{3}}\right)\right)dk \tag{6.223}$$

for $Y = O(1)$ as $t \to \infty$. We first consider the approximation of $F(Y,t)$ in (6.223) for $Y = O(1)(< 0)$ as $t \to \infty$. For $Y < 0$, the phase becomes stationary in (6.223) at $k = k_s(Y,t)$, and, via (6.13), we have

$$k_s(Y,t) = \frac{\sqrt{2}(-Y)^{\frac{1}{2}}}{t^{\frac{1}{3}}} + O\left(\frac{(-Y)^{\frac{3}{2}}}{t}\right)$$

for $Y = O(1)(< 0)$ as $t \to \infty$. Thus we may approximate $F(Y,t)$ as

$$F(Y,t) \sim \int_0^{k_s(Y,t)+\delta(t)} \frac{1}{k} \sin t\left(\gamma(k) - k\left(1 + Yt^{-\frac{2}{3}}\right)\right) dk \qquad (6.224)$$

for $Y = O(1)(< 0)$ and $\delta(t) = o(1)$ as $t \to \infty$. For $|k| \ll 1$ we have

$$\gamma(k) - k\left(1 - Yt^{-\frac{2}{3}}\right) = -\frac{1}{6}k^3 - Yt^{-\frac{2}{3}}k + O\left(k^5\right)$$

so that, as $k_s(Y,t) + \delta(t) = o(1)$ as $t \to \infty$, (6.224) becomes

$$F(Y,t) \sim -\int_0^{k_s(Y,t)+\delta(t)} \frac{1}{k} \sin t\left(\frac{1}{6}k^3 + Yt^{-\frac{2}{3}}k + O\left(k^5\right)\right) dk \qquad (6.225)$$

for $Y = O(1)(< 0)$ as $t \to \infty$. At this stage it is convenient to write

$$\frac{1}{k} \sin t\left(\frac{1}{6}k^3 + Yt^{-\frac{2}{3}}k + O\left(k^5\right)\right) = t^{\frac{1}{3}} \int_{X=0}^{Y} \cos t\left(\frac{1}{6}k^3 + Xt^{-\frac{2}{3}}k + O\left(k^5\right)\right) dX$$

$$+ \frac{1}{k} \sin\left(\frac{1}{6}tk^3 + O\left(k^5\right)\right)$$

$$(6.226)$$

so that, via (6.225) and (6.226), we have

$$F(Y,t) \sim -t^{\frac{1}{3}} \int_{X=0}^{Y} \int_{k=0}^{k_s(Y,t)+\delta(t)} \cos t\left(\frac{1}{6}k^3 + Xt^{-\frac{2}{3}}k + O\left(k^5\right)\right) dk\,dX$$

$$(6.227)$$

$$- \int_{k=0}^{k_s(Y,t)+\delta(t)} \frac{1}{k} \sin\left(\frac{1}{6}tk^3 + O\left(k^5\right)\right) dk$$

for $Y = O(1)(< 0)$ as $t \to \infty$. We first consider the second integral on the right-hand side of (6.227). Set

$$\hat{F}_1(Y,t) = \int_{k=0}^{k_s(Y,t)+\delta(t)} \frac{1}{k} \sin\left(\frac{1}{6}tk^3 + O\left(k^5\right)\right) dk \qquad (6.228)$$

for $Y = O(1)(< 0)$ as $t \to \infty$. Upon making the substitution

$$u = \frac{1}{6}tk^3$$

in (6.228), we obtain, on taking $\delta(t) = t^{-\frac{1}{4}}$,

$$\hat{F}_1(Y,t) \backsim \frac{1}{3} \int_0^\infty \frac{1}{u} \sin u \, du$$

for $Y = O(1)(<0)$ as $t \to \infty$. Hence

$$\hat{F}_1(Y,t) \backsim \frac{\pi}{6} \qquad (6.229)$$

for $Y = O(1)(<0)$ as $t \to \infty$. We now consider the first integral on the right-hand side of (6.227). Set

$$\hat{F}_2(Y,t) = \int_{X=0}^Y \int_{k=0}^{k_s(Y,t)+\delta(t)} \cos t \left(\frac{1}{6}k^3 + Xt^{-\frac{2}{3}}k + O\left(k^5\right) \right) dk\,dX \qquad (6.230)$$

for $Y = O(1)(<0)$ as $t \to \infty$. Upon making the substitution

$$v = \frac{k}{k_s(X,t)} - 1$$

in (6.230), we obtain, with $\delta(t) = t^{-\frac{1}{4}}$,

$$\hat{F}_2(Y,t) \backsim \int_{X=0}^Y \frac{\sqrt{2}(-X)^{\frac{1}{2}}}{t^{\frac{1}{3}}} \int_{v=-1}^\infty \cos \left(\left(2^{\frac{1}{3}}(-X)\right)^{\frac{3}{2}} \left(\frac{1}{3}v^3 + v^2 - \frac{2}{3} \right) \right) dv\,dX \qquad (6.231)$$

for $Y = O(1)(<0)$ as $t \to \infty$. From [Oliver *et al.* (2010)] (Chap. 9, p. 196) we have

$$Ai\left(2^{\frac{1}{3}}X\right) = \frac{\left(2^{\frac{1}{3}}(-X)\right)^{\frac{1}{2}}}{\pi} \int_{v=-1}^\infty \cos \left(\left(2^{\frac{1}{3}}(-X)\right)^{\frac{3}{2}} \left(\frac{1}{3}v^3 + v^2 - \frac{2}{3} \right) \right) dv \qquad (6.232)$$

for $X < 0$. Thus, via (6.231) and (6.232), we have

$$\hat{F}_2(Y,t) \backsim \frac{2^{\frac{1}{3}}\pi}{t^{\frac{1}{3}}} \int_{X=0}^Y Ai\left(2^{\frac{1}{3}}X\right) dX \qquad (6.233)$$

for $Y = O(1)(<0)$ as $t \to \infty$. It is now convenient to make the substitution

$$s = 2^{\frac{1}{3}}X$$

in (6.233) and then use the identity

$$\int_{z=-\infty}^0 Ai(z)\,dz = \frac{2}{3} \qquad (6.234)$$

([Oliver *et al.* (2010)] (Chap. 9, p. 202)), so that we finally have

$$\hat{F}_2(Y,t) \backsim -\frac{2\pi}{3t^{\frac{1}{3}}} + \frac{\pi}{t^{\frac{1}{3}}} \int_{s=-\infty}^{2^{\frac{1}{3}}Y} Ai(s)\,ds \qquad (6.235)$$

for $Y = O(1)(< 0)$ as $t \to \infty$. Thus, via (6.227), (6.228), (6.229), (6.230) and (6.235), we have

$$F(Y,t) \backsim \frac{\pi}{2} - \pi \int_{s=-\infty}^{2^{\frac{1}{3}}Y} Ai(s)\, ds \qquad (6.236)$$

for $Y = O(1)(< 0)$ as $t \to \infty$. We now consider the estimation of $F(Y,t)$ for $Y = O(1)(> 0)$ as $t \to \infty$. We first observe that

$$t^{\frac{1}{3}} \int_{X=0}^{Y} \cos t \left(\gamma(k) - k \left(1 + Xt^{-\frac{2}{3}} \right) \right) dX = -\frac{1}{k} \sin t \left(\gamma(k) - k \left(1 + Yt^{-\frac{2}{3}} \right) \right)$$

$$+ \frac{1}{k} \sin t \, (\gamma(k) - k) \qquad (6.237)$$

so that, via (6.223) and (6.237), we have

$$F(Y,t) = \int_{k=0}^{\infty} \frac{1}{k} \sin t \, (\gamma(k) - k) \, dk$$

$$\qquad (6.238)$$

$$- t^{\frac{1}{3}} \int_{k=0}^{\infty} \left(\int_{X=0}^{Y} \cos t \left(\gamma(k) - k \left(1 + Xt^{-\frac{2}{3}} \right) \right) dX \right) dk$$

for $Y = O(1)(> 0)$ and as $t \in \mathbb{R}^+$. We consider the first integral on the right-hand side of (6.238). Set

$$\hat{F}_3(t) = \int_{k=0}^{\infty} \frac{1}{k} \sin t \, (\gamma(k) - k) \, dk \qquad (6.239)$$

for $t \in \mathbb{R}^+$. The phase becomes stationary in (6.239) at $k = 0$, so as $t \to \infty$ we may approximate $\hat{F}_3(t)$ by

$$\hat{F}_3(t) \backsim \int_{k=0}^{\delta(t)} \frac{1}{k} \sin t \, (\gamma(k) - k) \, dk \qquad (6.240)$$

as $t \to \infty$ with $\delta(t) = o(1)$ as $t \to \infty$. For $|k| \ll 1$ we have

$$\gamma(k) - k = -\frac{1}{6} k^3 + O\left(k^5 \right)$$

so that (6.240) becomes

$$\hat{F}_3(t) \backsim -\int_{k=0}^{\delta(t)} \frac{1}{k} \sin \left(\frac{1}{6} tk^3 + O\left(tk^5 \right) \right) dk \qquad (6.241)$$

as $t \to \infty$. Upon making the substitution

$$u = \frac{1}{6} tk^3$$

in (6.241) we obtain, on taking $\delta(t) = t^{-\frac{1}{4}}$,

$$\hat{F}_3(t) \backsim -\frac{1}{3} \int_{u=0}^{\infty} \frac{1}{u} \sin u \, du$$

as $t \to \infty$. Hence

$$\hat{F}_3(t) \backsim -\frac{\pi}{6} \tag{6.242}$$

as $t \to \infty$. We now consider the second integral on the right-hand side of (6.238). Set

$$\hat{F}_4(Y,t) = t^{\frac{1}{3}} \int_{k=0}^{\infty} \left(\int_{X=0}^{Y} \cos t \left(\gamma(k) - k \left(1 + Xt^{-\frac{2}{3}} \right) \right) dX \right) dk \tag{6.243}$$

for $Y = O(1)(>0)$ with $t \in \mathbb{R}^+$. In order to estimate $\hat{F}_4(Y,t)$ for $Y = O(1)(>0)$ as $t \to \infty$ we first write

$$\left(\hat{F}_4(Y,t) \right)_R = t^{\frac{1}{3}} \int_{k=0}^{R} \left(\int_{X=0}^{Y} \cos t \left(\gamma(k) - k \left(1 + Xt^{-\frac{2}{3}} \right) \right) dX \right) dk \tag{6.244}$$

for $Y = O(1)(>0)$ with $t \in \mathbb{R}^+$ and $R > 0$. Upon interchanging the order of integration in (6.244) we may write

$$\left(\hat{F}_4(Y,t) \right)_R = t^{\frac{1}{3}} \int_{X=0}^{Y} \left(\int_{k=0}^{R} \cos t \left(\gamma(k) - k \left(1 + Xt^{-\frac{2}{3}} \right) \right) dk \right) dX \tag{6.245}$$

$$= \frac{t^{\frac{1}{3}}}{2} \left(\left(\hat{F}_+(Y,t) \right)_R + \left(\hat{F}_-(Y,t) \right)_R \right)$$

for $Y = O(1)(>0)$ with $t \in \mathbb{R}^+$ and $R > 0$, where

$$\left(\hat{F}_+(Y,t) \right)_R = \int_{X=0}^{Y} \left(\int_{k=0}^{R} \exp \left(it \left(\gamma(k) - k \left(1 + Xt^{-\frac{2}{3}} \right) \right) \right) dk \right) dX \tag{6.246}$$

and

$$\left(\hat{F}_-(Y,t) \right)_R = \int_{X=0}^{Y} \left(\int_{k=0}^{R} \exp \left(-it \left(\gamma(k) - k \left(1 + Xt^{-\frac{2}{3}} \right) \right) \right) dk \right) dX. \tag{6.247}$$

We consider first (6.246). The phase becomes stationary in the inner integral in (6.246) at $k = \pm i \tau_s(X,t)$ for $X \in [0,Y]$ with $t \in \mathbb{R}^+$. Hence, on taking $R > \frac{\pi}{2}$, we will deform the contour of integration in the inner integral in (6.246) from $k \in [0,R]$ onto the steepest descents contour D_-, shown in Fig. (6.20). To achieve this we consider the contour $C_{R_-} = [0,R] \cup D_2 \cup D_{R_-} \cup D_1$, where D_2 is an arc on the circle $|k| = R$, and D_{R_-} is a finite section of D_-, as shown in Fig. (6.39). The point $k = k_1$ is the intersection point of D_2 and D_{R_-}, and we recall, via (6.35), that

$$\theta_R = \frac{\pi}{2} + O \left(\frac{1}{R^{\frac{1}{2}}} \right)$$

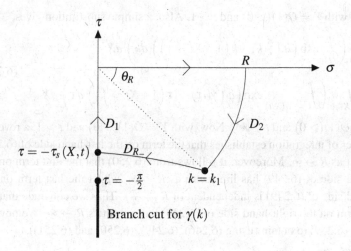

Fig. 6.39: Contour C_{R_-}.

as $R \to \infty$, where θ_R is the angle shown in Fig. (6.39). Now, via Cauchy's theorem, we have

$$\int_{C_{R_-}} \exp\left(it\left(\gamma(k) - k\left(1 + Xt^{-\frac{2}{3}}\right)\right)\right) dk = 0 \qquad (6.248)$$

for $X \in [0,Y]$ with $Y = O(1)(>0)$, $t \in \mathbb{R}^+$ and $R > \frac{\pi}{2}$. Therefore, via (6.248), we have

$$\int_{X=0}^{Y} \left(\int_{k=0}^{R} \exp\left(it\left(\gamma(k) - k\left(1 + Xt^{-\frac{2}{3}}\right)\right)\right) dk\right) dX$$

$$= -\int_{X=0}^{Y} \left(\int_{D_2} \exp\left(it\left(\gamma(k) - k\left(1 + Xt^{-\frac{2}{3}}\right)\right)\right) dk\right) dX$$

$$\qquad (6.249)$$

$$-\int_{X=0}^{Y} \left(\int_{D_{R_-}} \exp\left(it\left(\gamma(k) - k\left(1 + Xt^{-\frac{2}{3}}\right)\right)\right) dk\right) dX$$

$$-\int_{X=0}^{Y} \left(\int_{D_1} \exp\left(it\left(\gamma(k) - k\left(1 + Xt^{-\frac{2}{3}}\right)\right)\right) dk\right) dX$$

for $Y = O(1)(>0)$, $t \in \mathbb{R}^+$ and $R > \frac{\pi}{2}$. After a change in the order of integration, we may follow the approach in (6.40)–(6.44) to establish that

$$\int_{X=0}^{Y} \left(\int_{D_2} \exp\left(it\left(\gamma(k) - k\left(1 + Xt^{-\frac{2}{3}}\right)\right)\right) dk\right) dX = O\left(\frac{1}{R}\right) \qquad (6.250)$$

as $R \to \infty$ with $Y = O(1)(>0)$ and $t > 1$. Also, a simple substitution gives,

$$\int_{X=0}^{Y} \left(\int_{D_1} \exp\left(it\left(\gamma(k) - k\left(1 + Xt^{-\frac{2}{3}}\right)\right)\right) dk \right) dX$$

$$= i \int_{X=0}^{Y} \left(\int_{\tau=-\tau_s(X,t)}^{0} \exp\left(it\left(\gamma(i\tau) - i\tau\left(1 + Xt^{-\frac{2}{3}}\right)\right)\right) d\tau \right) dX \tag{6.251}$$

with $Y = O(1)(>0)$ and $t \in \mathbb{R}^+$. Now, with $Y = O(1)(>0)$ and $t > 1$, a reversal in the order of integration establishes that the term on the left-hand side of (6.249) has a limit as $R \to \infty$. Moreover, it follows from (6.250) that the first term on the right-hand side of (6.249) has limit zero as $R \to \infty$, whilst the last term on the right-hand side of (6.249) is independent of R ($> \frac{\pi}{2}$). Thus we conclude that the second term on the right-hand side of (6.249) has a limit as $R \to \infty$. We now let $R \to \infty$ in (6.249) to obtain (using (6.246), (6.249), (6.250) and (6.251)),

$$\lim_{R \to \infty} \left(\left(\hat{F}_+(Y,t) \right)_R \right)$$

$$= -\lim_{R \to \infty} \left(\int_{X=0}^{Y} \left(\int_{D_{R_-}} \exp\left(it\left(\gamma(k) - k\left(1 + Xt^{-\frac{2}{3}}\right)\right)\right) dk \right) dX \right) \tag{6.252}$$

$$- i \int_{X=0}^{Y} \left(\int_{\tau=-\tau_s(X,t)}^{0} \exp\left(it\left(\gamma(i\tau) - i\tau\left(1 + Xt^{-\frac{2}{3}}\right)\right)\right) d\tau \right) dX$$

for $Y = O(1)(>0)$ with $t > 1$. Similarly we obtain,

$$\lim_{R \to \infty} \left(\left(\hat{F}_-(Y,t) \right)_R \right)$$

$$= -\lim_{R \to \infty} \left(\int_{X=0}^{Y} \left(\int_{D_{R_+}} \exp\left(-it\left(\gamma(k) - k\left(1 + Xt^{-\frac{2}{3}}\right)\right)\right) dk \right) dX \right) \tag{6.253}$$

$$+ i \int_{X=0}^{Y} \left(\int_{\tau=-\tau_s(X,t)}^{0} \exp\left(it\left(\gamma(i\tau) - i\tau\left(1 + Xt^{-\frac{2}{3}}\right)\right)\right) d\tau \right) dX$$

for $Y = O(1)(>0)$ and $t > 1$, with D_{R_+} being the contour in the k-plane which is a reflection of D_{R_-} in the real k-axis. It now follows from (6.243)–(6.247), with (6.252) and (6.253), that,

$$\hat{F}_4(Y,t) = -\frac{1}{2}t^{\frac{1}{3}} \left(\lim_{R \to \infty} \left(\int_{X=0}^{Y} \left(\int_{D_{R_-}} \exp\left(it\left(\gamma(k) - k\left(1 + Xt^{-\frac{2}{3}}\right)\right)\right) dk \right) dX \right) \right.$$

$$\left. + \lim_{R \to \infty} \left(\int_{X=0}^{Y} \left(\int_{D_{R_+}} \exp\left(-it\left(\gamma(k) - k\left(1 + Xt^{-\frac{2}{3}}\right)\right)\right) dk \right) dX \right) \right) \tag{6.254}$$

for $Y = O(1)(>0)$ and $t > 1$. Now the inner integrals in (6.254), on D_{R_-} and D_{R_+}, are uniformly (and exponentially) convergent with respect to $X \in [0,Y]$ as $R \to \infty$. It then follows that

$$\lim_{R \to \infty} \left(\int_{X=0}^{Y} \left(\int_{D_{R_-}} \exp\left(it\left(\gamma(k) - k\left(1 + Xt^{-\frac{2}{3}}\right)\right)\right) dk \right) dX \right)$$

$$= -\int_{X=0}^{Y} \left(\int_{D_-} \exp\left(it\left(\gamma(k) - k\left(1 + Xt^{-\frac{2}{3}}\right)\right)\right) dk \right) dX$$

and

$$\lim_{R \to \infty} \left(\int_{X=0}^{Y} \left(\int_{D_{R_+}} \exp\left(-it\left(\gamma(k) - k\left(1 + Xt^{-\frac{2}{3}}\right)\right)\right) dk \right) dX \right)$$

$$= -\int_{X=0}^{Y} \left(\int_{D_+} \exp\left(-it\left(\gamma(k) - k\left(1 + Xt^{-\frac{2}{3}}\right)\right)\right) dk \right) dX$$

for $Y = O(1)(>0)$ and $t > 1$. Thus,

$$\hat{F}_4(Y,t) = \frac{1}{2} t^{\frac{1}{3}} \left(\int_{X=0}^{Y} \left(\int_{D_-} \exp\left(it\left(\gamma(k) - k\left(1 + Xt^{-\frac{2}{3}}\right)\right)\right) dk \right) dX \right. \tag{6.255}$$

$$\left. + \int_{X=0}^{Y} \left(\int_{D_+} \exp\left(-it\left(\gamma(k) - k\left(1 + Xt^{-\frac{2}{3}}\right)\right)\right) dk \right) dX \right)$$

for $Y = O(1)(>0)$ and $t > 1$. We now estimate the first integral (along D_-) on the right-hand side of (6.255) as $t \to \infty$ with $Y = O(1)(>0)$. The inner integral is in the form of a steepest descents integral as $t \to \infty$. Thus we estimate the inner integral near the stationary point at $k = -i\tau_s(X,t)$ (with $X \in [0,Y]$) as $t \to \infty$, where

$$\tau_s(X,t) = \frac{\sqrt{2}X^{\frac{1}{2}}}{t^{\frac{1}{3}}} + O\left(\frac{X^{\frac{3}{2}}}{t}\right) \tag{6.256}$$

for $X \in [0,Y]$ as $t \to \infty$. Therefore,

$$\int_{X=0}^{Y} \left(\int_{D_-} \exp\left(it\left(\gamma(k) - k\left(1 + Xt^{-\frac{2}{3}}\right)\right)\right) dk \right) dX$$

$$\sim \int_{X=0}^{Y} \left(\int_{\sigma=0}^{\delta(t)} \exp\left(it\left(\gamma(\sigma - i\tau_s(X,t)) - (\sigma - i\tau_s(X,t))\left(1 + Xt^{-\frac{2}{3}}\right)\right)\right) d\sigma \right) dX \tag{6.257}$$

for $Y = O(1)(>0)$ as $t \to \infty$, where we have set $k = \sigma - i\tau_s(X,t)$, with $\sigma \in [0, \delta(t))$, on the contour D_- close to $k = -i\tau_s(X,t)$, and $\delta(t) = o(1)$ as $t \to \infty$. For $|k| \ll 1$ we have

$$\gamma(k) - k\left(1 - Xt^{-\frac{2}{3}}\right) = -\frac{1}{6}k^3 - Xt^{-\frac{2}{3}}k + O\left(k^5\right) \tag{6.258}$$

so that, via (6.256) and (6.258), (6.257) becomes

$$\int_{X=0}^{Y} \left(\int_{D_-} \exp\left(it \left(\gamma(k) - k \left(1 + X t^{-\frac{2}{3}} \right) \right) \right) dk \right) dX$$

$$\sim \int_{X=0}^{Y} \exp\left(-\frac{2\sqrt{2}}{3} X^{\frac{3}{2}} \right) \left(\int_{\sigma=0}^{\delta(t)} \exp\left(-\frac{1}{6} it\sigma^3 - \frac{\sqrt{2}}{2} X^{\frac{1}{2}} t^{\frac{2}{3}} \sigma^2 \right) d\sigma \right) dX$$

$$(6.259)$$

for $Y = O(1)(>0)$ as $t \to \infty$, with $\delta(t) = o\left(t^{-\frac{1}{5}} \right)$ as $t \to \infty$. Upon making the substitution

$$u = \frac{t^{\frac{1}{3}}}{2^{\frac{1}{3}}} \sigma$$

in (6.259), we obtain, on taking $\delta(t) = O\left(t^{-\frac{1}{4}} \right)$

$$\int_{X=0}^{Y} \left(\int_{D_-} \exp\left(it \left(\gamma(k) - k \left(1 + X t^{-\frac{2}{3}} \right) \right) \right) dk \right) dX$$

$$\sim \frac{2^{\frac{1}{3}}}{t^{\frac{1}{3}}} \int_{X=0}^{Y} \exp\left(-\frac{2}{3} \left(2^{\frac{1}{3}} X \right)^{\frac{3}{2}} \right) \left(\int_{u=0}^{\infty} \exp\left(-\frac{1}{3} iu^3 - \left(2^{\frac{1}{3}} X \right)^{\frac{1}{2}} u^2 \right) du \right) dX$$

$$(6.260)$$

for $Y = O(1)(>0)$ as $t \to \infty$. Similarly, we may approximate the second term on the right-hand side of (6.255) as

$$\int_{X=0}^{Y} \left(\int_{D_+} \exp\left(-it \left(\gamma(k) - k \left(1 + X t^{-\frac{2}{3}} \right) \right) \right) dk \right) dX$$

$$\sim \frac{2^{\frac{1}{3}}}{t^{\frac{1}{3}}} \int_{X=0}^{Y} \exp\left(-\frac{2}{3} \left(2^{\frac{1}{3}} X \right)^{\frac{3}{2}} \right) \left(\int_{u=0}^{\infty} \exp\left(\frac{1}{3} iu^3 - \left(2^{\frac{1}{3}} X \right)^{\frac{1}{2}} u^2 \right) du \right) dX$$

$$(6.261)$$

for $Y = O(1)(>0)$ as $t \to \infty$. Thus, via (6.255), (6.260) and (6.261), we have

$$\hat{F}_4(Y,t) \sim 2^{\frac{1}{3}} \int_{X=0}^{Y} \exp\left(-\frac{2}{3} \left(2^{\frac{1}{3}} X \right)^{\frac{3}{2}} \right)$$

$$\times \left(\int_{u=0}^{\infty} \exp\left(-\left(2^{\frac{1}{3}} X \right)^{\frac{1}{2}} u^2 \right) \cos\left(\frac{1}{3} u^3 \right) du \right) dX$$

$$(6.262)$$

for $Y = O(1)(>0)$ as $t \to \infty$. The inner integral in (6.262) is in the form of an Airy function, [Oliver *et al.* (2010)] (Chap. 9, p. 196). Hence

$$\hat{F}_4(Y,t) \sim 2^{\frac{1}{3}} \pi \int_{X=0}^{Y} Ai\left(2^{\frac{1}{3}} X \right) dX$$

$$(6.263)$$

for $Y = O(1)(> 0)$ as $t \to \infty$. Therefore, via (6.238), (6.239), (6.242), (6.243) and (6.263), we have

$$F(Y,t) \sim -\frac{\pi}{6} - 2^{\frac{1}{3}}\pi \int_{X=0}^{Y} Ai\left(2^{\frac{1}{3}}X\right) dX \tag{6.264}$$

for $Y = O(1)(> 0)$ as $t \to \infty$. It is now convenient to let

$$s = 2^{\frac{1}{3}}X$$

in (6.264) and use (6.234) so that we finally have

$$F(Y,t) \sim \frac{\pi}{2} - \pi \int_{s=-\infty}^{2^{\frac{1}{3}}Y} Ai(s)\, ds \tag{6.265}$$

for $Y = O(1)(> 0)$ as $t \to \infty$. Thus, via (6.236) and (6.265), we have

$$F(Y,t) \sim \frac{\pi}{2} - \pi \int_{s=-\infty}^{2^{\frac{1}{3}}Y} Ai(s)\, ds \tag{6.266}$$

for $Y = O(1)$ as $t \to \infty$. In particular, we have, by [Oliver *et al.* (2010)] (Chap. 9, p. 202),

$$F(Y,t) \sim \frac{\pi}{2} - \frac{\pi^{\frac{1}{2}}}{2^{\frac{1}{4}}t^{\frac{1}{2}}(-Y)^{\frac{3}{4}}} \sin\left(\frac{\pi}{4} - \frac{2\sqrt{2}}{3}(-Y)^{\frac{3}{2}}\right) \tag{6.267}$$

for $(-Y) \gg 1$ as $t \to \infty$, and

$$F(Y,t) \sim -\frac{\pi}{2} + \frac{\pi^{\frac{1}{2}}}{2^{\frac{5}{4}}Y^{\frac{3}{4}}} \exp\left(-\frac{2\sqrt{2}}{3}Y^{\frac{3}{2}}\right) \tag{6.268}$$

for $Y \gg 1$ as $t \to \infty$. We observe that (6.267) (as $Y \to -\infty$) and (6.268) (as $Y \to +\infty$) asymptotically match accordingly with (6.125) and (6.156) respectively.

In a similar manner we may approximate $F(-y,t)$ in Region $\hat{1}^{-}$, where $y = -1 + \hat{Y}t^{-\frac{2}{3}}$, as

$$F(\hat{Y},t) \sim \frac{\pi}{2} - \pi \int_{s=-\infty}^{-2^{\frac{1}{3}}\hat{Y}} Ai(s)\, ds \tag{6.269}$$

for $\hat{Y} = O(1)$ as $t \to \infty$. In particular,

$$F(\hat{Y},t) \sim \frac{\pi}{2} - \frac{\sqrt{\pi}}{2^{\frac{1}{4}}\hat{Y}^{\frac{3}{4}}} \sin\left(\frac{\pi}{4} - \frac{2\sqrt{2}}{3}\hat{Y}^{\frac{3}{2}}\right) \tag{6.270}$$

for $\hat{Y} \gg 1$ as $t \to \infty$, and

$$F(\hat{Y},t) \sim -\frac{\pi}{2} + \frac{\sqrt{\pi}}{2^{\frac{5}{4}}\left(-\hat{Y}\right)^{\frac{3}{4}}} \exp\left(-\frac{2\sqrt{2}}{3}\left(-\hat{Y}\right)^{\frac{3}{2}}\right) \tag{6.271}$$

for $\left(-\hat{Y}\right) \gg 1$ as $t \to \infty$.

6.7   Inner Region Coordinate Expansion for $\bar{\eta}(y,t)$ as $t \to \infty$

We now have, via (6.5), (6.217), (6.218), (6.219), (6.220), (6.221), (6.222), (6.266), (6.267), (6.268), (6.269), (6.270) and (6.271), the following approximations for $\bar{\eta}(y,t)$ in the following regions,

Region $\hat{\mathbf{I}}^+$: $y\left(=\frac{x}{t}\right) = 1 + Yt^{-\frac{2}{3}}$, with $Y = O(1)$ as $t \to \infty$. We have,

$$\bar{\eta}(Y,t) \backsim \frac{1}{2}\left(1 - \int_{s=-\infty}^{2^{\frac{1}{3}}Y} Ai(s)\,ds\right) + \frac{2^{\frac{1}{3}}\beta}{4t^{\frac{1}{3}}}Ai\left(2^{\frac{1}{3}}Y\right) \tag{6.272}$$

for $Y = O(1)$ as $t \to \infty$. In particular,

$$\bar{\eta}(Y,t) \backsim \frac{1}{2^{\frac{9}{4}}\sqrt{\pi}}\left(\frac{1}{Y^{\frac{3}{4}}} + \frac{\beta}{\sqrt{2}t^{\frac{1}{3}}Y^{\frac{1}{4}}}\right)\exp\left(-\frac{2\sqrt{2}}{3}Y^{\frac{3}{2}}\right) \tag{6.273}$$

for $Y \gg 1$ as $t \to \infty$, and

$$\bar{\eta}(Y,t) \backsim \frac{1}{2} - \frac{1}{2^{\frac{5}{4}}\sqrt{\pi}}\left(\frac{1}{(-Y)^{\frac{3}{4}}}\sin\left(\frac{\pi}{4} - \frac{2\sqrt{2}}{3}(-Y)^{\frac{3}{2}}\right)\right.$$
$$\left. - \frac{\beta}{\sqrt{2}t^{\frac{1}{3}}(-Y)^{\frac{1}{4}}}\cos\left(\frac{\pi}{4} - \frac{2\sqrt{2}}{3}(-Y)^{\frac{3}{2}}\right)\right) \tag{6.274}$$

for $(-Y) \gg 1$ as $t \to \infty$, and it is readily verified that (6.273) and (6.274) asymptotically match with (6.196) and (6.194) respectively according to the asymptotic matching principal of Van Dyke [Van Dyke (1975)].

Region $\hat{\mathbf{I}}^-$: $y\left(=\frac{x}{t}\right) = -1 + \hat{Y}t^{-\frac{2}{3}}$, with $\hat{Y} = O(1)$ as $t \to \infty$. We have,

$$\bar{\eta}\left(\hat{Y},t\right) \backsim \frac{1}{2}\left(1 + \int_{s=-\infty}^{-2^{\frac{1}{3}}\hat{Y}} Ai(s)\,ds\right) + \frac{2^{\frac{1}{3}}\beta}{4t^{\frac{1}{3}}}Ai\left(-2^{\frac{1}{3}}\hat{Y}\right) \tag{6.275}$$

for $\hat{Y} = O(1)$ as $t \to \infty$. In particular,

$$\bar{\eta}\left(\hat{Y},t\right) \backsim \frac{1}{2} + \frac{1}{2^{\frac{5}{4}}\sqrt{\pi}}\left(\frac{1}{\hat{Y}^{\frac{3}{4}}}\sin\left(\frac{\pi}{4} - \frac{2\sqrt{2}}{3}\hat{Y}^{\frac{3}{2}}\right) + \frac{\beta}{\sqrt{2}t^{\frac{1}{3}}\hat{Y}^{\frac{1}{4}}}\cos\left(\frac{\pi}{4} - \frac{2\sqrt{2}}{3}\hat{Y}^{\frac{3}{2}}\right)\right) \tag{6.276}$$

for $\hat{Y} \gg 1$ as $t \to \infty$, and

$$\bar{\eta}\left(\hat{Y},t\right) \backsim 1 - \frac{1}{2^{\frac{9}{4}}\sqrt{\pi}}\left(\frac{1}{\left(-\hat{Y}\right)^{\frac{3}{4}}} - \frac{\beta}{\sqrt{2}t^{\frac{1}{3}}\left(-\hat{Y}\right)^{\frac{1}{4}}}\right)\exp\left(-\frac{2\sqrt{2}}{3}\left(-\hat{Y}\right)^{\frac{3}{2}}\right) \tag{6.277}$$

for $\left(-\widehat{Y}\right) \gg 1$ as $t \to \infty$, and it is readily verified that (6.276) and (6.277) asymptotically match with (6.201) and (6.203) respectively according to the asymptotic matching principal of Van Dyke [Van Dyke (1975)].

As an illustration, graphs of $\bar{\eta}(y,t)$ in each region (as determined from (6.272)–(6.277)) are given in Figs. (6.40) and (6.41), where we have set $t = 300$ and $\beta = 1$.

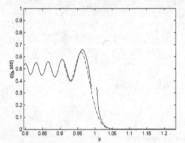

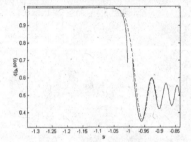

Fig. 6.40: The graph of $\bar{\eta}(y,300)$ in Region $\hat{I}^{+}$, (6.272), ($--$), and in Region I^{+}, (6.191) and Region II^{+} (6.195), ($-$).

Fig. 6.41: The graph of $\bar{\eta}(y,300)$ in Region $\hat{I}^{-}$, (6.275), ($--$), and in Region I^{-}, (6.198) and Region II^{-} (6.202), ($-$).

Chapter 7

Summary of the Asymptotic Structure of $\bar{\eta}(x,t)$ as $t \to 0$ and $t \to \infty$

In this chapter we give a summary of the exact solution for the free surface displacement $\bar{\eta}(x,t)$ for $(x,t) \in \mathbb{R} \times \mathbb{R}^+$ and the asymptotic structure of $\bar{\eta}(x,t)$ as $t \to 0$ and $t \to \infty$, as determined in Chap. 3, Chap. 4 and Chap. 6 respectively. Illustrations of the asymptotic structure are given in each case and graphs of the approximations in each case are also shown.

7.1 Exact Solution for $\bar{\eta}(x,t)$

We have now completed the asymptotic structure to the free surface solution $\bar{\eta}(x,t)$ as $t \to 0$ and as $t \to \infty$. We recall that the free surface displacement $\bar{\eta}(x,t)$ is given exactly by

$$\bar{\eta}(x,t) = \frac{1}{2\pi\beta} \int_{C_\delta} \frac{1}{k^2} (1 - \exp(i\beta k)) \cos(\gamma(k)t) \exp(-ixk) \, dk, \qquad (7.1)$$

for $(x,t) \in \mathbb{R} \times \bar{\mathbb{R}}^+$, where

$$\gamma(k) = (k \tanh k)^{\frac{1}{2}}$$

and C_δ is the contour along on the real k-axis, indented below the origin by a semi-circle of radius $0 < \delta \ll 1$, as shown in Fig. (7.1). We have investigated coordinate expansions for $\bar{\eta}(x,t)$, as given by (7.1), and have obtained uniform asymptotic approximations for $\bar{\eta}(x,t)$ in the limits $t \to 0$ and $t \to \infty$ for $x \in \mathbb{R}$. The asymptotic structure of $\bar{\eta}(x,t)$ in each of these cases is summarised below.

7.2 Asymptotic Structure of $\bar{\eta}(x,t)$ as $t \to 0$

Let $N_0(t)$ and $N_\beta(t)$ be $O(t^2)$ neighbourhoods of the points $x = 0$ and $x = \beta$ as $t \to 0$. Then we have the following outer region asymptotic approximation for $\bar{\eta}(x,t)$,

125

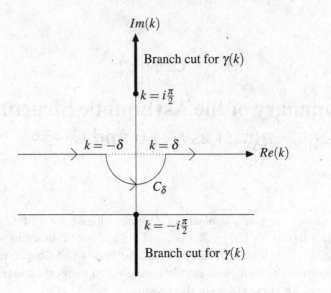

Fig. 7.1: Contour C_δ in the k-plane.

Outer Region $x \in \mathbb{R} \setminus \{N_0(t) \cup N_\beta(t)\}$ as $t \to 0$.
In the outer region,

$$\bar{\eta}(x,t) = \begin{cases} \dfrac{t^2}{2\pi\beta} \log\left(\dfrac{\tanh\frac{\pi}{4}x}{\tanh\frac{\pi}{4}(x-\beta)}\right) + o(t^2), & \text{as } t \to 0 \text{ with} \\[4mm] & x \in [\beta, \infty) \setminus N_\beta(t), \\[4mm] \dfrac{1}{\beta}(\beta - x) + \dfrac{t^2}{2\pi\beta} \log\left(\dfrac{\tanh\frac{\pi}{4}x}{\tanh\frac{\pi}{4}(\beta - x)}\right) + o(t^2), & \text{as } t \to 0 \text{ with} \\[4mm] & x \in [0, \beta] \setminus \{N_0(t) \cup N_\beta(t)\}, \\[4mm] 1 + \dfrac{t^2}{2\pi\beta} \log\left(\dfrac{\tanh\frac{\pi}{4}(-x)}{\tanh\frac{\pi}{4}(\beta - x)}\right) + o(t^2), & \text{as } t \to 0 \text{ with} \\[4mm] & x \in (-\infty, 0] \setminus N_0(t). \end{cases}$$

(7.2)

In particular, in the outer region we have

$$\bar{\eta}(x,t) = \frac{t^2}{2\pi\beta}\left(-\log(x - \beta) + \log\left(\frac{4}{\pi}\tanh\left(\frac{\pi}{4}\beta\right)\right) + O\left((x-\beta)^2\right)\right) + o\left(t^2\right)$$

when $0 < (x - \beta) \ll 1$ as $t \to 0$.

$$\bar{\eta}(x,t) = \frac{1}{\beta}(\beta - x) + \frac{t^2}{2\pi\beta}\left(-\log(\beta - x) + \log\left(\frac{4}{\pi}\tanh\left(\frac{\pi}{4}\beta\right)\right)\right) + O\left((x - \beta)^2\right)$$

$$+ o\left(t^2\right)$$

$$(7.3)$$

when $0 < (\beta - x) \ll 1$ as $t \to 0$.

$$\bar{\eta}(x,t) = \frac{1}{\beta}(\beta - x) + \frac{t^2}{2\pi\beta}\left(\log(x) - \log\left(\frac{4}{\pi}\tanh\left(\frac{\pi}{4}\beta\right)\right) + O\left(x^2\right)\right) + o\left(t^2\right)$$

when $0 < x \ll 1$ as $t \to 0$.

$$\bar{\eta}(x,t) = 1 + \frac{t^2}{2\pi\beta}\left(\log(-x) - \log\left(\frac{4}{\pi}\tanh\left(\frac{\pi}{4}\beta\right)\right) + O\left(x^2\right)\right) + o\left(t^2\right)$$

when $0 < (-x) \ll 1$ as $t \to 0$.

$$\bar{\eta}(x,t) = \frac{t^2}{2\pi\beta}\left(2\left(\exp\left(\frac{\pi}{2}\beta\right) - 1\right)\exp\left(-\frac{\pi}{2}x\right) + O\left(\exp\left(-\frac{3\pi}{2}x\right)\right)\right) + o\left(t^2\right)$$

when $x \gg 1$ as $t \to 0$.

$$\bar{\eta}(x,t) = 1 + \frac{t^2}{2\pi\beta}\left(2\left(\exp\left(-\frac{\pi}{2}\beta\right) - 1\right)\exp\left(\frac{\pi}{2}x\right) + O\left(\exp\left(\frac{3\pi}{2}x\right)\right)\right) + o\left(t^2\right)$$

when $(-x) \gg 1$ as $t \to 0$.

We have the following inner region approximations for $\bar{\eta}(x,t)$:

Inner Region A $x \in N_0(t)$ as $t \to 0$.
In inner region A,

$$\bar{\eta}(X,t) = 1 + \frac{t^2}{\pi\beta}\left(\log t + H(X) - \pi X - \frac{1}{2}\log\left(\tanh\frac{\pi}{4}\beta\right)\right) + o(t^2) \qquad (7.4)$$

for $X = O(1)$ as $t \to 0$, with $x = t^2 X$.

Inner Region B $x \in N_\beta(t)$ as $t \to 0$.
In inner region B,

$$\bar{\eta}(\bar{X},t) = \frac{t^2}{\pi\beta}\left(-\log t - H(\bar{X}) + \frac{1}{2}\log\left(\tanh\frac{\pi}{4}\beta\right)\right) + o(t^2) \qquad (7.5)$$

for $\bar{X} = O(1)$ as $t \to 0$, with $x = \beta + \bar{X}t^2$.

In (7.4) and (7.5)

$$H(X) = \begin{cases} F_1(X) + F_2(X) + F_3(X) - \dfrac{1}{2}c_4, & X \geq 0 \\[2mm] F_1(-X) + F_2(-X) + F_3(-X) + \pi X - \dfrac{1}{2}c_4, & X < 0, \end{cases}$$

with

$$F_1(X) = \int_{s=1}^{\infty} \frac{1}{s^2} \left(\cos\left(s^{\frac{1}{2}}\right) - 1 \right) \cos(Xs)\, ds,$$

$$F_2(X) = \int_{s=0}^{1} \frac{1}{2s} \left(1 - \cos(Xs) \right) ds,$$

$$F_3(X) = \int_{s=0}^{1} h(s) \cos(Xs)\, ds,$$

where

$$h(s) = \begin{cases} \dfrac{1}{2s}\left(1 + \dfrac{2}{s}\left(\cos\left(s^{\frac{1}{2}}\right) - 1\right)\right), & s > 0, \\[3mm] \dfrac{1}{24}, & s = 0. \end{cases}$$

and

$$c_4 = \int_{k=0}^{1} \frac{\tanh k}{k}\, dk + \int_{k=1}^{\infty} \frac{\tanh k - 1}{k}\, dk = 0.818780\ldots.$$

In particular, in Inner Region A we have

$$\bar{\eta}(X,t) = 1 + \frac{t^2}{\pi\beta}\left(\log t + \left(c_1 + c_3 - \frac{1}{2}c_4 - 1\right) - \frac{1}{2}\log\left(\tanh\left(\frac{\pi}{4}\beta\right)\right)\right.$$

$$\left. -\frac{1}{2}\pi X + o\left(|X|^{\frac{3}{2}}\right)\right) + o(t^2)$$

when $|X| \ll 1$ as $t \to 0$, whilst

$$\bar{\eta}(X,t) = 1 + \frac{t^2}{\pi\beta}\left(-\pi X + \log t + \frac{1}{2}\log\left(\frac{\pi}{4}X\right) - \frac{1}{2}\log\left(\tanh\left(\frac{\pi}{4}\beta\right)\right) + O\left(\frac{1}{X^2}\right)\right)$$

$$+ o(t^2)$$

when $X \gg 1$ as $t \to 0$, and

$$\bar{\eta}(X,t) = 1 + \frac{t^2}{\pi\beta}\left(\log t + \frac{1}{2}\log\left(-\frac{\pi}{4}X\right) - \frac{1}{2}\log\left(\tanh\left(\frac{\pi}{4}\beta\right)\right) + O\left(\frac{1}{X^2}\right)\right)$$

$$+ o(t^2)$$

when $(-X) \gg 1$ as $t \to 0$. In Inner Region B we have

$$\bar{\eta}(\bar{X},t) = \frac{t^2}{\pi\beta} \left(-\log t - \left(c_1 + c_3 - \frac{1}{2}c_4 - 1 \right) + \frac{1}{2}\log\left(\tanh\left(\frac{\pi}{4}\beta\right)\right) \right.$$

$$\left. -\frac{1}{2}\pi\bar{X} + o\left(|\bar{X}|^{\frac{3}{2}}\right) \right) + o\left(t^2\right)$$

when $|\bar{X}| \ll 1$ as $t \to 0$, whilst

$$\bar{\eta}(\bar{X},t) = \frac{t^2}{\pi\beta} \left(-\log t - \frac{1}{2}\log\left(\frac{\pi}{4}\bar{X}\right) + \frac{1}{2}\log\left(\tanh\left(\frac{\pi}{4}\beta\right)\right) + O\left(\frac{1}{\bar{X}^2}\right) \right)$$

$$+ o\left(t^2\right)$$

when $\bar{X} \gg 1$ as $t \to 0$, and

$$\bar{\eta}(\bar{X},t) = \frac{t^2}{\pi\beta} \left(-\pi\bar{X} - \log t - \frac{1}{2}\log\left(-\frac{\pi}{4}\bar{X}\right) + \frac{1}{2}\log\left(\tanh\left(\frac{\pi}{4}\beta\right)\right) + O\left(\frac{1}{\bar{X}^2}\right) \right)$$

$$+ o\left(t^2\right)$$

when $(-\bar{X}) \gg 1$ as $t \to 0$, where

$$c_1 = \int_1^\infty \frac{1}{s^2}\cos(s^{\frac{1}{2}})\,ds = 0.036242\ldots$$

and

$$c_3 = \int_0^1 h(s)\,ds = 0.040980\ldots.$$

An illustration of the asymptotic structure of $\bar{\eta}(x,t)$ as $t \to 0$, with $x \in \mathbb{R}$, is given in Fig. (7.2). With $\beta = 1$, graphs of $\bar{\eta}(x,t)$ against x are shown in Fig. (7.3) for $t \in [0,0.5]$. Details close to inner regions A and B are shown in Fig. (7.4). In Fig. (7.4) we note that close to the initial corners at $x = \beta$ and $x = 0$, the structure of $\bar{\eta}(x,t)$ as $t \to 0$ shows incipient localised jet formation close to $x = \beta$ (in inner region B) and incipient localised collapse close to $x = 0$ (in inner region A).

7.3   Asymptotic Structure of $\bar{\eta}(x,t)$ as $t \to \infty$

We have the asymptotic approximation to $\bar{\eta}(x,t)$ as $t \to \infty$ in the following outer regions:

Region II$^+$: $y \left(= \frac{x}{t}\right) > 1 + o(1)$ as $t \to \infty$. We have,

$$\bar{\eta}(y,t) \sim \frac{1}{2\beta\sqrt{2\pi t v_{kk}(-i\tau_s(y))}\tau_s(y)^2} \left(\exp(\beta\tau_s(y)) - 1\right)\exp\left(-tv\left(-i\tau_s(y),y\right)\right)$$

$$\tag{7.6}$$

as $t \to \infty$, where $k = \pm i\tau_s(y)$ (with $\tau_s(y) > 0$) are the stationary points of $\gamma(k) - yk$ when $y > 1 + o(1)$, and $v(k,y) = Im(\gamma(k) - ky)$. In particular,

$$\bar{\eta}(y,t) \backsim \frac{1}{2^{\frac{9}{4}}\pi^{\frac{1}{2}}t^{\frac{1}{2}}} \left(\frac{1}{(y-1)^{\frac{3}{4}}} + \frac{\beta}{\sqrt{2}(y-1)^{\frac{1}{4}}} \right) \exp\left(-\frac{2\sqrt{2}}{3}t(y-1)^{\frac{3}{2}} \right) \quad (7.7)$$

when $0 < (y-1) \ll 1$ as $t \to \infty$, and

$$\bar{\eta}(y,t) \backsim \frac{1}{2\pi\beta} \left(\exp\left(\frac{\pi}{2}\beta\right) - 1 \right) \left(\frac{4}{\sqrt{6}\pi^{\frac{4}{3}}t^{\frac{1}{2}}y^{\frac{5}{6}}} \right) \exp\left(-\frac{\pi}{2}t\left(y - \frac{3}{\pi^{\frac{2}{3}}}y^{\frac{1}{3}} \right) \right)$$

$$(7.8)$$

when $y \gg 1$ as $t \to \infty$.

Region I$^+$: $o(1) < y\left(= \frac{x}{t}\right) < 1 - o(1)$ as $t \to \infty$. We have,

$$\bar{\eta}(y,t) \backsim \frac{1}{2} + \frac{1}{\beta\sqrt{2\pi t\gamma''(-k_s(y))}k_s(y)^2} \left(\cos\left(\frac{\pi}{4} + tk_s(y)\,(y - c(k_s(y))) \right) \right.$$

$$(7.9)$$

$$\left. - \cos\left(\frac{\pi}{4} - \beta k_s(y) + tk_s(y)\,(y - c(k_s(y))) \right) \right) \right) \quad \text{as } t \to \infty,$$

with

$$c(k) = \frac{\gamma(k)}{k} \quad (7.10)$$

and $k = k_s(y)$ is the positive stationary point of $\gamma(k) - yk$ when $o(1) < y < 1 - o(1)$. In particular,

$$\bar{\eta}(y,t) \backsim \frac{1}{2} + \frac{8y^{\frac{5}{2}}}{\beta\pi^{\frac{1}{2}}t^{\frac{1}{2}}} \left(\cos\left(\frac{\pi}{4} - \frac{t}{4y} \right) - \cos\left(\frac{\pi}{4} - \frac{\beta}{4y^2} - \frac{t}{4y} \right) \right) \quad (7.11)$$

when $0 < y \ll 1$ as $t \to \infty$, and

$$\bar{\eta}(y,t) \backsim \frac{1}{2} - \frac{1}{2^{\frac{5}{4}}\pi^{\frac{1}{2}}t^{\frac{1}{2}}} \left(\frac{1}{(1-y)^{\frac{3}{4}}} \sin\left(\frac{\pi}{4} - \frac{2\sqrt{2}}{3}t(1-y)^{\frac{3}{2}} \right) \right.$$

$$(7.12)$$

$$\left. - \frac{\beta}{\sqrt{2}(1-y)^{\frac{1}{4}}} \cos\left(\frac{\pi}{4} - \frac{2\sqrt{2}}{3}t(1-y)^{\frac{3}{2}} \right) \right)$$

when $0 < (1-y) \ll 1$ as $t \to \infty$.

Region I$^-$: $-1 + o(1) < y\left(= \frac{x}{t}\right) < -o(1)$ as $t \to \infty$. We have,

$$\bar{\eta}(y,t) \backsim \frac{1}{2} + \frac{1}{\beta\sqrt{2\pi t\gamma''(-k_s(-y))}k_s(-y)^2} \left(\cos\left(-\frac{\pi}{4} + tk_s(-y)\,(y + c(k_s(-y))) \right) \right.$$

$$\left. - \cos\left(-\frac{\pi}{4} - \beta k_s(-y) + tk_s(-y)\,(y + c(k_s(-y))) \right) \right) \right) \quad \text{as } t \to \infty,$$

$$(7.13)$$

with

$$c(k) = \frac{\gamma(k)}{k} \tag{7.14}$$

where $k = k_s(-y)$ is the positive stationary point of $\gamma(k) + yk$ when $-1 + o(1) < y < -o(1)$. In particular,

$$\bar{\eta}(y,t) \backsim \frac{1}{2} + \frac{8(-y)^{\frac{5}{2}}}{\beta \pi^{\frac{1}{2}} t^{\frac{1}{2}}} \left(\cos\left(\frac{\pi}{4} + \frac{t}{4y}\right) - \cos\left(\frac{\pi}{4} + \frac{\beta}{4y^2} + \frac{t}{4y}\right) \right) \tag{7.15}$$

when $0 < (-y) \ll 1$ as $t \to \infty$, and

$$\bar{\eta}(y,t) \backsim \frac{1}{2} + \frac{1}{2^{\frac{5}{4}}\pi^{\frac{1}{2}}t^{\frac{1}{2}}} \left(\frac{1}{(1+y)^{\frac{3}{4}}} \sin\left(\frac{\pi}{4} - \frac{2\sqrt{2}}{3}t(1+y)^{\frac{3}{2}}\right) \right.$$

$$\left. + \frac{\beta}{\sqrt{2}(1+y)^{\frac{1}{4}}} \cos\left(\frac{\pi}{4} - \frac{2\sqrt{2}}{3}t(1+y)^{\frac{3}{2}}\right) \right) \tag{7.16}$$

when $0 < (1+y) \ll 1$ as $t \to \infty$.

Region II$^-$: $y\left(= \frac{x}{t}\right) < -1 - o(1)$ as $t \to \infty$. We have,

$$\bar{\eta}(y,t) \backsim 1 - \frac{1}{2\beta\sqrt{2\pi t v_{kk}(-i\tau_s(-y))}\tau_s(-y)^2} (1 - \exp(\beta\tau_s(-y)))$$

$$\times \exp(-tv(-i\tau_s(-y),y)), \tag{7.17}$$

as $t \to \infty$, where $k = \pm i\tau_s(-y)$ (with $\tau_s(-y) > 0$) are the stationary points of $\gamma(k) + yk$ when $y < -1 - o(1)$. In particular,

$$\bar{\eta}(y,t) \backsim 1 - \frac{1}{2^{\frac{9}{4}}\pi^{\frac{1}{2}}t^{\frac{1}{2}}} \left(\frac{1}{(-(y+1))^{\frac{3}{4}}} - \frac{\beta}{\sqrt{2}(-(y+1))^{\frac{1}{4}}} \right)$$

$$\times \exp\left(-\frac{2\sqrt{2}}{3}t(-(y+1))^{\frac{3}{2}}\right) \tag{7.18}$$

when $0 < (-(y+1)) \ll 1$ as $t \to \infty$, and

$$\bar{\eta}(y,t) \backsim 1 - \frac{1}{2\pi\beta} \left(1 - \exp\left(-\frac{\pi}{2}\beta\right)\right) \left(\frac{4}{\sqrt{6}\pi^{\frac{4}{3}}t^{\frac{1}{2}}(-y)^{\frac{5}{6}}}\right) \exp\left(\frac{\pi}{2}t\left(y - \frac{3}{\pi^{\frac{2}{3}}}y^{\frac{1}{3}}\right)\right) \tag{7.19}$$

when $(-y) \gg 1$ as $t \to \infty$.

We now have the asymptotic approximation to $\bar{\eta}(x,t)$ as $t \to \infty$ in the following inner regions:

Region $\hat{\mathbf{I}}^+$: $y\left(=\frac{x}{t}\right) = 1 + Yt^{-\frac{2}{3}}$, with $Y = O(1)$ as $t \to \infty$. We have,

$$\bar{\eta}(Y,t) \backsim \frac{1}{2}\left(1 - \int_{s=-\infty}^{2^{\frac{1}{3}}Y} Ai(s)\,ds\right) + \frac{2^{\frac{1}{3}}\beta}{4t^{\frac{1}{3}}}Ai\left(2^{\frac{1}{3}}Y\right) \quad \text{as } t \to \infty. \qquad (7.20)$$

In particular,

$$\bar{\eta}(Y,t) \backsim \frac{1}{2^{\frac{9}{4}}\sqrt{\pi}}\left(\frac{1}{Y^{\frac{3}{4}}} + \frac{\beta}{\sqrt{2}t^{\frac{1}{3}}Y^{\frac{1}{4}}}\right)\exp\left(-\frac{2\sqrt{2}}{3}Y^{\frac{3}{2}}\right) \qquad (7.21)$$

for $Y \gg 1$ as $t \to \infty$, and

$$\bar{\eta}(Y,t) \backsim \frac{1}{2} - \frac{1}{2^{\frac{5}{4}}\sqrt{\pi}}\left(\frac{1}{(-Y)^{\frac{3}{4}}}\sin\left(\frac{\pi}{4} - \frac{2\sqrt{2}}{3}(-Y)^{\frac{3}{2}}\right)\right.$$
$$\left. - \frac{\beta}{\sqrt{2}t^{\frac{1}{3}}(-Y)^{\frac{1}{4}}}\cos\left(\frac{\pi}{4} - \frac{2\sqrt{2}}{3}(-Y)^{\frac{3}{2}}\right)\right) \qquad (7.22)$$

for $(-Y) \gg 1$ as $t \to \infty$.

Region $\hat{\mathbf{I}}^-$: $y\left(=\frac{x}{t}\right) = -1 + \hat{Y}t^{-\frac{2}{3}}$, with $\hat{Y} = O(1)$ as $t \to \infty$. We have,

$$\bar{\eta}\left(\hat{Y},t\right) \backsim \frac{1}{2}\left(1 + \int_{s=-\infty}^{-2^{\frac{1}{3}}\hat{Y}} Ai(s)\,ds\right) + \frac{2^{\frac{1}{3}}\beta}{4t^{\frac{1}{3}}}Ai\left(-2^{\frac{1}{3}}\hat{Y}\right) \quad \text{as } t \to \infty. \quad (7.23)$$

In particular,

$$\bar{\eta}\left(\hat{Y},t\right) \backsim \frac{1}{2} + \frac{1}{2^{\frac{5}{4}}\sqrt{\pi}}\left(\frac{1}{\hat{Y}^{\frac{3}{4}}}\sin\left(\frac{\pi}{4} - \frac{2\sqrt{2}}{3}\hat{Y}^{\frac{3}{2}}\right) + \frac{\beta}{\sqrt{2}t^{\frac{1}{3}}\hat{Y}^{\frac{1}{4}}}\cos\left(\frac{\pi}{4} - \frac{2\sqrt{2}}{3}\hat{Y}^{\frac{3}{2}}\right)\right) \qquad (7.24)$$

for $\hat{Y} \gg 1$ as $t \to \infty$, and

$$\bar{\eta}\left(\hat{Y},t\right) \backsim 1 - \frac{1}{2^{\frac{9}{4}}\sqrt{\pi}}\left(\frac{1}{\left(-\hat{Y}\right)^{\frac{3}{4}}} - \frac{\beta}{\sqrt{2}t^{\frac{1}{3}}\left(-\hat{Y}\right)^{\frac{1}{4}}}\right)\exp\left(-\frac{2\sqrt{2}}{3}\left(-\hat{Y}\right)^{\frac{3}{2}}\right) \qquad (7.25)$$

for $\left(-\hat{Y}\right) \gg 1$ as $t \to \infty$.

An illustration of the asymptotic structure for $\bar{\eta}(y,t)$ as $t \to \infty$, with $y \in \mathbb{R}$, is shown in Fig. (7.5). With $\beta = 1$, a graph of $\bar{\eta}(y,t)$ against y, with $t = 70$, is shown in Fig. (7.6). Details close to the inner regions $\hat{\mathbf{I}}^+$ and $\hat{\mathbf{I}}^-$ are shown in Figs. (7.7)–(7.10). In Figs. (7.5) and (7.6) the structure of $\bar{\eta}(y,t)$ shows oscillatory behaviour for $-1+o(1) < y < 1-o(1)$, where we note that there is a weak singularity in our approximation to $\bar{\eta}(y,t)$ (in (7.9) and (7.13)) at $y = 0$. However we observe from (7.11) and (7.15) that $\bar{\eta}(y,t) \to \frac{1}{2}$ as $y \to 0$. This oscillatory region is connected to regions, with $y > 1+o(1)$ and $y < -1-o(1)$, where we observe an exponentially small disturbance to the far field conditions. These regions are connected by localised inner regions, where $y = 1 \pm O\left(t^{-\frac{2}{3}}\right)$ and $y = -1 \pm O\left(t^{-\frac{2}{3}}\right)$, and graphs of these regions, as determined from (7.20)–(7.25), are given in Figs. (7.7) and (7.8) for $t = 70$, and in Figs. (7.9) and (7.10) for $t = 300$.

In the next chapter we consider the numerical evaluation of the exact form of $\bar{\eta}(x,t)$, as given by (7.1).

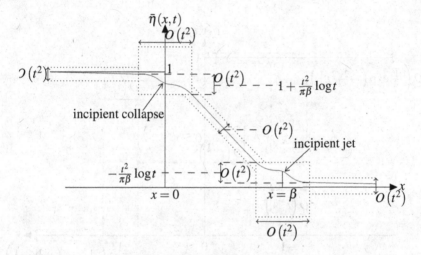

Fig. 7.2: A sketch for the asymptotic structure of $\bar{\eta}(x,t)$ as $t \to 0$.

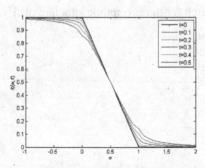

Fig. 7.3: Graph of $\bar{\eta}(x,t)$ with $\beta = 1$ and $t \in [0,0.5]$.

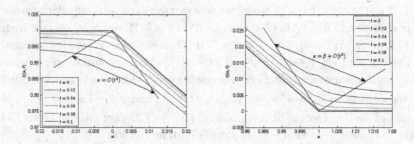

Fig. 7.4: Graphs of $\bar{\eta}(x,t)$, with $\beta = 1$, in inner region A and inner region B for $t \in [0, 0.1]$, illustrating the incipient localised collapse and jet structure respectively.

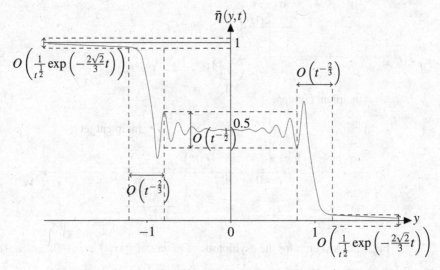

Fig. 7.5: A sketch for the asymptotic structure of $\bar{\eta}(y,t)$ as $t \to \infty$.

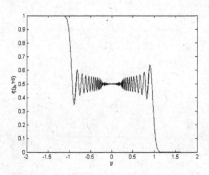

Fig. 7.6: Graph of $\bar{\eta}(y,t)$ for $t = 70$.

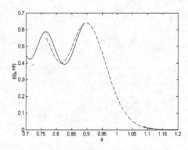

Fig. 7.7: The graph of $\bar{\eta}(y,70)$ in Region $\hat{\mathrm{I}}^{+}$, (7.20), $(--)$, in Region I^{+}, (7.9) and Region II^{+} (7.6), $(-)$.

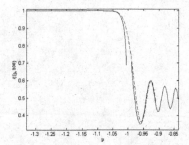

Fig. 7.8: The graph of $\bar{\eta}(y,70)$ in Region $\hat{\mathrm{I}}^{-}$, (7.23), $(--)$, in Region I^{-}, (7.13) and Region II^{-} (7.17), $(-)$.

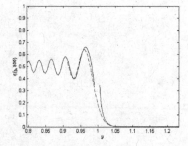

Fig. 7.9: The graph of $\bar{\eta}(y,300)$ in Region $\hat{\mathrm{I}}^{+}$, (7.20), $(--)$, in Region I^{+}, (7.9) and Region II^{+} (7.6), $(-)$.

Fig. 7.10: The graph of $\bar{\eta}(y,300)$ in Region $\hat{\mathrm{I}}^{-}$, (7.23), $(--)$, in Region I^{-}, (7.13) and Region II^{-} (7.17), $(-)$.

Chapter 8

Numerical Evaluation of the Exact Form of $\bar{\eta}(x,t)$

In this chapter we give a numerical evaluation of the exact free surface solution. We apply Simpson's rule to $\bar{\eta}(x,t)$ and graph the numerical evaluations, which show excellent agreement with the asymptotic structure summarised in Chap. 7. In order to efficiently numerically evaluate $\bar{\eta}(x,t)$, we recall that

$$\bar{\eta}(x,t) = \frac{1}{2\pi\beta} \left(I(x,t) - I(x-\beta,t) \right),$$

for $(x,t) \in \mathbb{R} \times \bar{\mathbb{R}}^+$, with

$$I(x,t) = \int_{C_\delta} \frac{1}{k^2} \cos(\gamma(k)t) \exp(-ikx)\, dk,$$

and

$$I(-x,t) = I(x,t) - 2\pi x \tag{8.1}$$

for $(x,t) \in \mathbb{R} \times \bar{\mathbb{R}}^+$, where

$$\gamma(k) = (k\tanh k)^{\frac{1}{2}}.$$

We will now numerically evaluate $I(x,t)$ for $(x,t) \in \bar{\mathbb{R}}^+ \times \bar{\mathbb{R}}^+$. For numerical evaluation, it is convenient to use (4.19), which gives

$$I(x,t) = 2 \int_0^\infty \frac{1}{k^2} \left(\cos\left(\gamma(k)t\right) - 1 \right) \cos(kx)\, dk \tag{8.2}$$

for $(x,t) \in \bar{\mathbb{R}}^+ \times \bar{\mathbb{R}}^+$. $\bar{\eta}(x,t)$ can then be constructed, via (3.32) and (3.34), as

$$\bar{\eta}(x,t) = \begin{cases} \dfrac{1}{2\pi\beta} \left(I(x,t) - I(x-\beta,t) \right), & \text{for } (x,t) \in [\beta,\infty) \times \bar{\mathbb{R}}^+, \\[3mm] \dfrac{1}{2\pi\beta} \left(2\pi(\beta-x) + I(x,t) - I(\beta-x,t) \right), & \text{for } (x,t) \in (0,\beta) \times \bar{\mathbb{R}}^+, \\[3mm] \dfrac{1}{2\pi\beta} \left(2\pi\beta + I(-x,t) - I(\beta-x,t) \right), & \text{for } (x,t) \in (-\infty,0] \times \bar{\mathbb{R}}^+. \end{cases}$$

Using (8.2), $I(x,t)$, for $(x,t) \in \bar{\mathbb{R}}^+ \times \mathbb{R}^+$, is numerically evaluated using Simpson's rule, after which we use (8.1) to determine $I(x,t)$ for $(x,t) \in \mathbb{R} \times \mathbb{R}^+$. Taking $\beta = 1$ and setting the upper limit of integration in (8.2) to be Δ, the error, E, associated with evaluating (8.2) via the composite Simpson's rule, satisfies the error bound

$$|E| \le \frac{\Delta^5}{2880m^4} \sup_{k \in [0,\infty)} \left| f^{(4)}(k,x,t) \right| \tag{8.3}$$

for each fixed x and t, (see [Süli and Mayers (2003)], p. 211) where

$$f(k,x,t) = \frac{1}{k^2} \left(\cos\left(\gamma(k)t\right) - 1 \right) \cos(kx)$$

for $k \in [0,\infty)$ and $(x,t) \in \times \bar{\mathbb{R}}^+ \times \mathbb{R}^+$, with m being the number of intervals the range of integration has been divided into, and the step length is

$$h = \frac{\Delta}{2m}.$$

A calculation of $\sup_{k \in [0,\infty)} \left| f^{(4)}(k,x,t) \right|$ gives the estimate

$$\sup_{k \in [0,\infty)} \left| f^{(4)}(k,x,t) \right| \le k_1 t^2 + k_2 t^4 + k_3 t^6 + k_4 x^2 t^2 + k_5 x^4 t^2 + k_6 x^2 t^4 \tag{8.4}$$

for all $(x,t) \in \times \bar{\mathbb{R}}^+ \times \mathbb{R}^+$, with k_i $(i = 1,\ldots,6)$ being positive numerical constants, independent of x and t. It then follows from (8.3) and (8.4) that

$$|E| \le \frac{\Delta^5}{2880m^4} \left(k_1 t^2 + k_2 t^4 + k_3 t^6 + k_4 x^2 t^2 + k_5 x^4 t^2 + k_6 x^2 t^4 \right)$$

for all $(x,t) \in \times \bar{\mathbb{R}}^+ \times \mathbb{R}^+$. Estimates of k_i $(i = 1,\ldots,6)$ are given by,

$$k_1 \le \frac{8}{5}, \quad k_2 \le \frac{2}{3}, \quad k_3 \le \frac{1}{30}, \quad k_4 \le 2, \quad k_5 \le \frac{1}{2}, \quad k_6 \le \frac{1}{2}.$$

In addition, the error induced by truncating the range of integration in (8.2) to $[0,\Delta]$ is given by T, where

$$|T| \le \frac{2}{\Delta}$$

for all $(x,t) \in \times \bar{\mathbb{R}}^+ \times \mathbb{R}^+$. Thus, the total error in approximating (8.2) via Simpson's rule is E_T, where

$$|E_T| \le |T| + |E|$$

$$\le \frac{2}{\Delta} + \frac{\Delta h^4}{180} \left(\frac{8}{5}t^2 + \frac{2}{3}t^4 + \frac{1}{30}t^6 + 2x^2 t^2 + \frac{1}{2}x^4 t^2 + \frac{1}{2}x^2 t^4 \right) \tag{8.5}$$

for all $(x,t) \in \times \bar{\mathbb{R}}^+ \times \mathbb{R}^+$. We can now control the error of the numerical approximation to (8.2) using (8.5).

All the subsequent computations performed in this chapter have choices of h and Δ such that $|E_T| \leq 10^{-4}$ for all $(x,t) \in \bar{\mathbb{R}}^+ \times \bar{\mathbb{R}}^+$ for which the computations are performed. Fig. (8.1) shows $\bar{\eta}(x,t)$ for $t \in [0,0.5]$, which has excellent agreement with Fig. (7.3). Inner regions A and B are shown in Figs. (8.2) and (8.3), which have excellent agreement with Fig. (7.4). Fig. (8.4) shows $\bar{\eta}(x,t)$ for $t \in [0.6, 1.2]$, and Figs. (8.5)–(8.8) show $\bar{\eta}(x,t)$ for $t \in [2, 16]$. Figs. (8.9) and (8.12) show $\bar{\eta}(y,t)$ for $t = 32$ and $t = 70$ respectively, where $y = \frac{x}{t}$. Region $\hat{I}^+$ and Region $\hat{I}^-$ (as defined in Sect. 7.3) for each case are shown in Figs. (8.10), (8.11), (8.13) and (8.14) and we observe excellent agreement between Figs. (8.12), (8.13), (8.14) and (7.6), (7.7), (7.8).

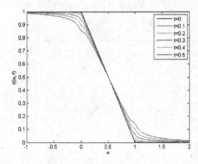

Fig. 8.1: The graph of $\bar{\eta}(x,t)$.

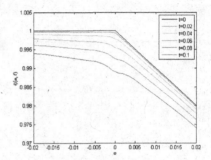

Fig. 8.2: The graph of $\bar{\eta}(x,t)$ in inner region A.

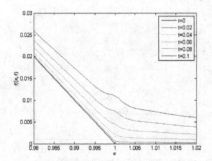

Fig. 8.3: The graph of $\bar{\eta}(x,t)$ in inner region B.

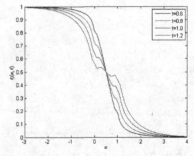

Fig. 8.4: The graph of $\bar{\eta}(x,t)$ for $t \in [0.6, 1.2]$.

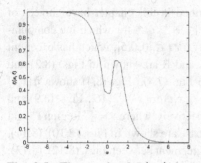

Fig. 8.5: The graph of $\bar{\eta}(x,t)$ for $x \in [-8,8]$ with $t = 2$.

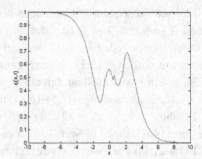

Fig. 8.6: The graph of $\bar{\eta}(x,t)$ for $x \in [-10,10]$ with $t = 4$.

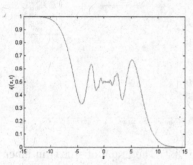

Fig. 8.7: The graph of $\bar{\eta}(x,t)$ for $x \in [-15,15]$ with $t = 8$.

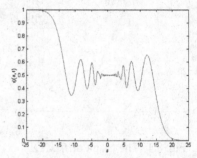

Fig. 8.8: The graph of $\bar{\eta}(x,t)$ for $x \in [-25,25]$ with $t = 16$.

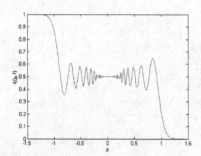

Fig. 8.9: The graph of $\bar{\eta}(y,t)$ for $y \in [-1.5,1.5]$ with $t = 32$.

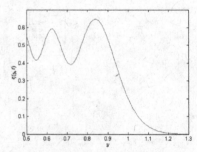

Fig. 8.10: The graph of $\bar{\eta}(x,t)$ for $t = 32$ in Region $\hat{I}^{+}$.

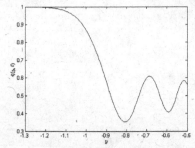

Fig. 8.11: The graph of $\bar{\eta}(x,t)$ for $t = 32$ in Region $\hat{I}^-$.

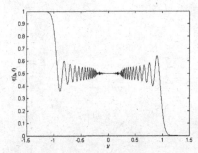

Fig. 8.12: The graph of $\bar{\eta}(y,t)$ for $y \in [-1.5, 1.5]$ with $t = 70$.

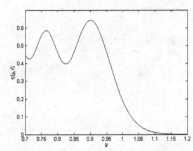

Fig. 8.13: The graph of $\bar{\eta}(x,t)$ for $t = 70$ in Region $\hat{I}^+$.

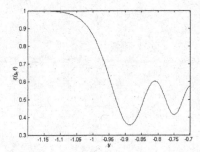

Fig. 8.14: The graph of $\bar{\eta}(x,t)$ for $t = 70$ in Region $\hat{I}^-$.

Chapter 9

Comparison with the Linearised Shallow Water Theory

In this chapter we consider the situation when the free surface displacement is modelled by the linearised shallow water theory. The free surface is governed by the one-dimensional wave equation, subject to initial conditions. We find the free surface solution in terms of D'Alemberts general solution and we graph the solutions for various times. Comparisons with results in Chaps. 7 and 8 show that the shallow water theory gives the general shape of the free surface, whilst remaining monotone and oscillation free.

9.1 Linearised Shallow Water Theory

We now consider the case when the free surface displacement, $\bar{\eta}(x,t)$, is described by the one-dimensional wave equation, that is

$$\bar{\eta}_{tt}(x,t) - \bar{\eta}_{xx}(x,t) = 0$$

for $(x,t) \in \mathbb{R} \times \mathbb{R}^+$, which is the governing equation for the free surface displacement for the linearised dam-break problem according to shallow water theory. The initial displacement of the free surface is given as

$$\bar{\eta}(x,0) = \bar{\eta}_0(x) = \begin{cases} 0, & x \geq \beta \\ \dfrac{1}{\beta}(\beta - x), & 0 < x < \beta \\ 1, & x \leq 0 \end{cases} \qquad (9.1)$$

and the free surface is initially at rest so we also have the initial condition

$$\bar{\eta}_t(x,0) = 0 \qquad (9.2)$$

for $x \in \mathbb{R}$. D'Alembert's general solution, together with initial conditions (9.1) and (9.2), then gives

$$\bar{\eta}(x,t) = \frac{1}{2}\bar{\eta}_0(x-t) + \frac{1}{2}\bar{\eta}_0(x+t) \qquad (9.3)$$

143

for $(x,t) \in \mathbb{R} \times \mathbb{R}^+$. With $\beta = 1$, graphs of $\bar{\eta}(x,t)$, as given in (9.3), for $t \in [0,1.2]$ are shown in Figs. (9.1) and (9.2). Graphs of $\bar{\eta}(x,t)$ for $t = 2,4,8,16$ are shown in Fig. (9.3), and graphs of $\bar{\eta}(y,t)$, where $y = \frac{x}{t}$, for $t \in [5,70]$ are shown in Figs. (9.4)–(9.6) respectively. Sketches of the structure of $\bar{\eta}(x,t)$ as $t \to 0^+$, and of $\bar{\eta}(y,t)$ as $t \to \infty$, for the linearised shallow water theory, as given by (9.3), are illustrated in Figs. (9.7) and (9.8) respectively. These are to be compared with the sketches of the structure of the solution to the full linearised theory, as given in Fig. (7.2) and Fig. (7.5). In fact, when $t \to 0^+$ we obtain from (9.1) and (9.3),

$$\bar{\eta}(x,t) = \begin{cases} 0, & x \geq \beta + t \\ \dfrac{1}{2\beta}(\beta - (x-t)), & \beta - t < x < \beta + t \\ \dfrac{1}{\beta}(\beta - x), & t \leq x \leq \beta - t \\ \dfrac{1}{2\beta}(2\beta - (x+t)), & -t < x < t \\ 1 & x \leq -t. \end{cases} \tag{9.4}$$

Similarly, when $t \to \infty$ we obtain from (9.1) and (9.3), on letting $y = \frac{x}{t}$,

$$\bar{\eta}(y,t) = \begin{cases} 0, & y \geq \dfrac{\beta}{t} + 1 \\ \dfrac{1}{2\beta}(\beta - t(y-1)), & 1 < y < \dfrac{\beta}{t} + 1 \\ \dfrac{1}{2}, & \dfrac{\beta}{t} - 1 \leq y \leq 1 \\ \dfrac{1}{2\beta}(2\beta - t(y+1)), & -1 < y < \dfrac{\beta}{t} - 1 \\ 1 & y \leq -1. \end{cases} \tag{9.5}$$

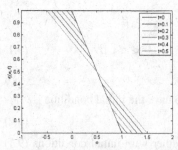

Fig. 9.1: The graph of $\bar{\eta}(x,t)$ for $t \in [0,0.5]$.

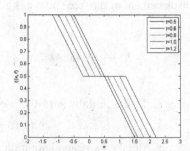

Fig. 9.2: The graph of $\bar{\eta}(x,t)$ for $t \in [0.5,1.2]$.

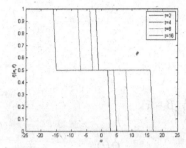

Fig. 9.3: The graph of $\bar{\eta}(x,t)$ for $t \in [2,16]$.

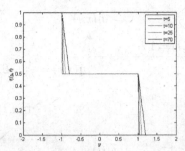

Fig. 9.4: The graph of $\bar{\eta}(y,t)$ for $t \in [5,70]$.

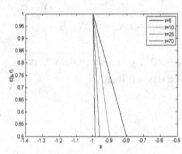

Fig. 9.5: The graph of $\bar{\eta}(y,t)$ for $t \in [5,70]$ with $y \in [-1.4,-0.5]$.

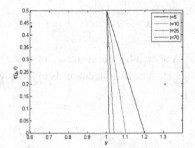

Fig. 9.6: The graph of $\bar{\eta}(y,t)$ for $t \in [5,70]$ with $y \in [0.6,1.4]$.

9.2   Comparison of the Linearised Shallow Water Theory with the Full Linearised Theory

We now compare (9.4) with (7.2), (7.4), (7.5) and (9.5) with (7.6)–(7.25) (as illustrated in Figs. (9.7), (7.2) and Figs. (9.8), (7.5)). As $t \to 0^+$, the linearised shallow water theory reproduces the leading order behaviour shown in the full linearised theory, but does not capture the incipient jet formation and collapse seen in the full linearised theory. In particular, in the inner regions of the full linearised theory, which are of thickness $O\left(t^2\right)$, we observe incipient jet formation and collapse, and the change in the initial displacement of the surface is $O\left(t^2 \log t\right)$. This is not reproduced in the shallow water theory, where instead we find the inner regions are of thickness $O(t)$, and the surface forms a linear slope, and the change in the initial displacement of the surface is $O(t)$. In the outer regions of the full linearised theory we observe an $O\left(t^2\right)$ change in the initial displacement of the

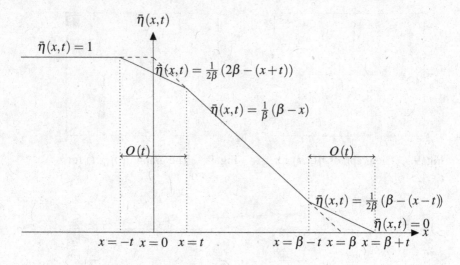

Fig. 9.7: A sketch for the structure of $\bar{\eta}(x,t)$ for $t \to 0^+$, given by (9.4), where the initial free surface displacement is also shown as a dashed line.

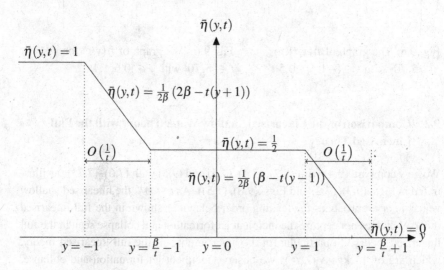

Fig. 9.8: A sketch for the structure of $\bar{\eta}(y,t)$ for $t \to \infty$ given by (9.5).

surface, compared to the linearised shallow water theory where there is no change in the initial displacement of the surface. Overall, the shallow water theory gives a good approximation to the full linearised theory as $t \to 0^+$, with the most notable difference being the structure of the surface in each of the inner regions. As $t \to \infty$, the linearised shallow water theory again reproduces the leading order behaviour shown in the full linearised theory, but does not capture the oscillatory behaviour. In particular, in the full linearised theory, the two outer regions, where $|y| > 1 \pm O\left(t^{-\frac{2}{3}}\right)$, show the surface differs from the initial conditions by an exponentially small order in t as $t \to \infty$. In the shallow water theory, these outer regions, where $|y| \geq 1 \pm O\left(t^{-1}\right)$, show no change in the initial conditions. Therefore, there is only an exponentially small difference between the two approximations in these regions. In the third outer region, the full linearised theory shows that the surface exhibits oscillatory behaviour about $\bar{\eta}(y,t) = \frac{1}{2}$, where the amplitude of the waves are $O\left(t^{-\frac{1}{2}}\right)$, whereas the linearised shallow water theory fails to capture these oscillations, and the surface remains constant at $\bar{\eta}(y,t) = \frac{1}{2}$. In the two inner regions, the full linearised theory shows the connection between the oscillating and exponentially decaying outer regions, and these inner regions are of thickness $O\left(t^{-\frac{2}{3}}\right)$. The linearised shallow water theory, however, has the surface as a linear slope in both inner regions, and the regions are of thickness $O\left(t^{-1}\right)$. There is an $O(1)$ difference in the structure of the surface in the inner regions. Overall, the linearised shallow water theory has good agreement with the full linearised theory in the outer regions where the difference between the two approximations becomes increasingly small as $t \to \infty$. The inner regions, where there is an $O(1)$ structural difference between the linearised shallow water and the full linearised approximations, become increasingly thinner as $t \to \infty$. Formally, a comparison of the asymptotic approximation as $t \to 0^+$ to the full linearised theory, ((7.2), (7.4) and (7.5)), with the solution to the linearised shallow water theory, (9.4), establishes that, with $\eta_E(x,t)$ and $\eta_{SW}(x,t)$ being the surface displacement according to the full linearised theory and the linearised shallow water theory respectively, then as $t \to 0^+$

$$\eta_E(x,t) - \eta_{SW}(x,t) = \begin{cases} O\left(t^2\right), & x \in (-\infty,-t) \cup (t,\beta-t) \cup \beta + t,\infty) \\ O(t), & x \in (-t,t) \cup (\beta-t,\beta+t). \end{cases}$$

Similarly, a comparison of the asymptotic approximation as $t \to \infty$ to the full linearised theory, ((7.6), (7.9), (7.13), (7.17), (7.20) and (7.23)), with the solution to the linearised shallow water theory (9.5) establishes that, as $t \to \infty$,

$$\eta_E(y,t) - \eta_{SW}(y,t) = \begin{cases} O\left(\dfrac{1}{t^{\frac{1}{2}}}\exp\left(-tv\left(-i\tau_s(y),y\right)\right)\right), & y \in (1+\Delta(t),\infty) \\[2ex] O(1), & y \in (-1-\Delta(t),-1+\Delta(t)) \\ & \quad \cup(1-\Delta(t),1+\Delta(t)) \\[1ex] O\left(t^{-\frac{1}{3}}\right), & y \in (-1+\Delta(t),1-\Delta(t)) \\[2ex] O\left(\dfrac{1}{t^{\frac{1}{2}}}\exp\left(-tv\left(-i\tau_s(-y),y\right)\right)\right), & y \in (-\infty,-1-\Delta(t)) \end{cases}$$

with $\Delta(t) = O\left(t^{-\frac{2}{3}}\right)$ as $t \to \infty$, and where $v\left(-i\tau_s(y),y\right) > 0$ for $y > 1$.

9.3 Comparison of the Linearised Shallow Water Theory with the Numerical Approximation

Comparisons between the numerical approximation to the full linearised theory obtained in Chap. 8 and the linearised shallow water theory are shown in Figs. (9.9)–(9.18). Graphs of $\bar{\eta}(x,t)$, computed numerically in Chap. 8, are shown with graphs of the linearised shallow water solution given by (9.3). We observe that the linearised shallow water solution improves as $t \to 0^+$ and $t \to \infty$.

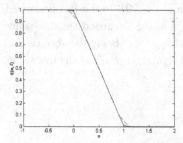
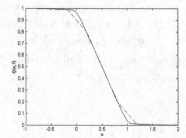

Fig. 9.9: The graph of $\bar{\eta}(x,t)$ for $t = 0.1$, computed numerically via Simpson's method ($-$) for the full linearised theory, and from the linearised shallow water theory (9.3) ($--$).

Fig. 9.10: The graph of $\bar{\eta}(x,t)$ for $t = 0.2$, computed numerically via Simpson's method ($-$), for the full linearised theory, and from the linearised shallow water theory (9.3) ($--$).

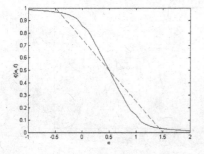

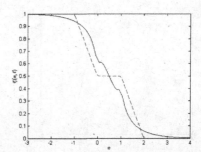

Fig. 9.11: The graph of $\bar{\eta}(x,t)$ for $t = 0.5$, computed numerically via Simpson's method $(-)$, for the full linearised theory, and from the linearised shallow water theory (9.3) $(--)$.

Fig. 9.12: The graph of $\bar{\eta}(x,t)$ for $t = 1$, computed numerically via Simpson's method $(-)$, for the full linearised theory, and from the linearised shallow water theory (9.3) $(--)$.

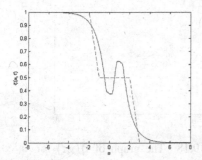

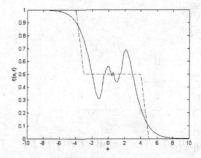

Fig. 9.13: The graph of $\bar{\eta}(x,t)$ for $t = 2$, computed numerically via Simpson's method $(-)$, for the full linearised theory, and from the linearised shallow water theory (9.3) $(--)$.

Fig. 9.14: The graph of $\bar{\eta}(x,t)$ for $t = 4$, computed numerically via Simpson's method $(-)$, for the full linearised theory, and from the linearised shallow water theory (9.3) $(--)$.

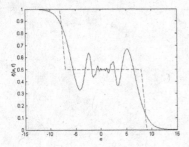

Fig. 9.15: The graph of $\bar{\eta}(x,t)$ for $t = 8$, computed numerically via Simpson's method $(-)$, for the full linearised theory, and from the linearised shallow water theory (9.3) $(--)$.

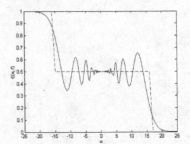

Fig. 9.16: The graph of $\bar{\eta}(x,t)$ for $t = 16$, computed numerically via Simpson's method $(-)$, for the full linearised theory, and from the linearised shallow water theory (9.3) $(--)$.

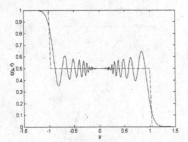

Fig. 9.17: The graph of $\bar{\eta}(y,t)$ for $t = 32$, computed numerically via Simpson's method $(-)$, for the full linearised theory, and from the linearised shallow water theory (9.3) $(--)$.

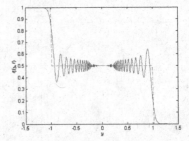

Fig. 9.18: The graph of $\bar{\eta}(y,t)$ for $t = 70$, computed numerically via Simpson's method $(-)$, for the full linearised theory, and from the linearised shallow water theory (9.3) $(--)$.

9.4 Conclusions

In this monograph we have considered a linearised fully two-dimensional dam-break problem where an inclined dam, with a small step height and slope, separates a horizontal layer of incompressible and inviscid fluid from a shallower horizontal layer of the fluid. The fluid was initially at rest on a horizontal flat, impermeable base and was bounded above by a free surface. The aim of the monograph was to solve the full linearised problem for the free surface and fluid velocity potential, and to obtain short time and large time asymptotic approximations of the free surface in detail via the approximation of Fourier-type integrals in the complex plane. Numerical approximation of the free surface for the full linearised theory was then compared with the asymptotic approximations. These results were also compared to the situation when the free surface was governed by the linearised shallow water theory. It has been found, significantly, that the full linearised theory and the linearised shallow water theory have good agreement in the short and large time in the outer asymptotic regions but differ at $O(1)$ in thin inner regions located at the upstream and downstream transition waterfronts.

In Chap. 2 we introduced a fluid velocity potential and formulated the non-linear dam-break problem via the conservation of mass and momentum with appropriate boundary and initial conditions. In Chap. 3 we considered the case of a dam with a small step height and slope resulting in the full linearised problem, which was solved via the complex Fourier transform to give an exact solution for the fluid velocity potential and free surface in terms of Fourier integrals in the complex plane. In Chap. 4 we obtained a uniform asymptotic approximation to the free surface during the initial stages of the flow. It was found that the asymptotic approximation was composed of three outer regions, which were connected by two inner regions, these being thin regions around the initial corners of the free surface displacement. The asymptotic approximation determined that the change in the surface displacement from the initial conditions was of $O\left(t^2\right)$ as $t \to 0$ in the outer regions, and revealed incipient jet formation and collapse in each inner region. The asymptotic approximation agreed with an experimental study by [Stansby *et al.* (1998)] which showed a jet formation near the initial position of the dam, and is also similar to the results in [Korobkin and Yilmaz (2009)] where jet formation was observed in an inner region around the point where the dam met the base. The asymptotic approximation is also in accord with that presented by Stoker, [Stoker (1957)], in that the change in the surface displacement from the initial condition is $O\left(t^2\right)$ as $t \to 0$ in the outer regions. In Chap. 5 we obtained a uniform asymptotic approximation to the free surface in the far field. The asymptotic approximation consisted of three regions and demonstrated that the free surface in the far field only changes from the initial displacement by an exponentially small order in x, as $|x| \to \infty$ in the far fields. In Chap. 6 we obtained a uniform asymptotic approximation to the free surface for large time. The asymptotic approximation consisted of an outer region in which the free surface exhibited algebraically decaying oscillatory behaviour. This connected to two inner regions which were described by Airy functions and their integrals, and each of these regions then connected to outer regions which extended into the far field, and in these outer regions the free surface differs from the initial conditions by an exponentially small order in t as $t \to \infty$. In Chap. 7 we gave a detailed summary of the uniform asymptotic approximations to the free surface for small and large times. In Chap. 8 a numerical approximation of the exact free surface solution was given. The exact solution was estimated via Simpson's rule and a precise error bound was given. The asymptotic approximations were in excellent agreement with the numerical approximations to the free surface for small and large times. In Chap. 9 we considered the situation where the free surface displacement is governed by the linearised shallow water theory, with the same initial surface displacement as the full linearised theory. The wave equation was solved by D'Alemberts general solution, which showed that according to the linearised shallow water theory, the free surface consisted of two travelling waves. A detailed comparison was then made between the solution of the full linearised theory and that of the linearised shallow water theory.

Bibliography

Ahmad, M., Mamat, M., Wan Nik, W., and Kartono, A. (2013). Numerical method for dam break problem by using Godunov approach, *Appl. Math. Comput.* **18**, 3, pp. 234–44.

Ancey, C., Iverson, R., Rentschler, M., and Denlinger, R. (2008). An exact solution for ideal dam-break floods on steep slopes, *Water Resour. Res.* **44**, pp. 1–10.

Brufau, P. and Garcia-Navarro, P. (2000). Two-dimensional dam break flow simulation, *Int. J. Numer. Meth. Fluid.* **33**, 1, pp. 35–57.

Dutykh, D. and Mitsotakis, D. (2010). On the relevance of the dam break problem in the context of nonlinear shallow water equations, *Discrete Cont. Dyn-S.* **3**, pp. 1–20.

Fernandez-Feria, R. (2006). Dam-break flow for arbitrary slopes of the bottom, *J. Eng. Math.* **54**, 4, pp. 319–331.

Goater, A. and Hogg, A. (2011). Bounded dam-break flows with tailwaters, *J. Fluid Mech.* **686**, pp. 160–186.

Hogg, A. (2006). Lock-release gravity currents and dam-break flows, *J. Fluid Mech.* **569**, pp. 61–87.

Hu, C. and Sueyoshi, M. (2010). Numerical simulation and experiment on dam break problem, *J. Marine Sci. Appl.* **9**, 2, pp. 109–114.

Hunt, B. (1983). Asymptotic solution for dam break on sloping channel, *J. Hydraul. Eng.* **109**, 12, pp. 1698–1706.

Korobkin, A. and Yilmaz, O. (2009). The initial stage of dam-break flow, *J. Eng. Math.* **63**, 2-4, pp. 293–308.

Leach, J. and Needham, D. (2008). The large-time development of the solution to an initial-value problem for the Korteweg–de Vries equation: I. Initial data has a discontinuous expansive step, *Nonlinearity* **21**, 10, p. 2391.

Oliver, F., Lozier, D., Boisvert, R., and Clark, C. (2010). *NIST Handbook of Mathematical Functions Hardback and CD–ROM.*

Ostapenko, V. (2003). Dam-break flows over a bottom step, *J. Appl. Me.ch. Tech. Phys.* **44**, 4, pp. 495–505.

Ostapenko, V. (2007). Modified shallow water equations which admit the propagation of discontinuous waves over a dry bed, *J. Appl. Mech. Tech. Phys.* **48**, 6, pp. 795–812.

Pohle, F. (2006). *The Lagrangian equations of hydrodynamics: solutions which are functions of time*, Ph.D. thesis, New York University.

Sopta, L., Maćešić, S., Holjević, D., Črnjarić-Žic, N., Škifić, J., Družeta, S., and Crnković, B. (2007). Tribalj dam-break and flood wave propagation, *Annali Univ Ferrara* **53**, 2, pp. 405–415.

Stansby, P., Chegini, A., and Barnes, T. (1998). The initial stages of dam-break flow, *J. Fluid Mech.* **374**, pp. 407–424.

Stoker, J. (1957). *Water waves*.

Süli, E. and Mayers, D. (2003). *An introduction to numerical analysis*.

Ungarish, M., Borden, Z., and Meiburg, E. (2014). Gravity currents with tailwaters in Boussinesq and non-Boussinesq systems: two-layer shallow-water dam-break solutions and Navier–Stokes simulations, *Environ. Fluid Mech.* **14**, 2, pp. 451–470.

Van Dyke, M. (1975). *Perturbation methods in fluid mechanics*.

via DIALOG., D. F. A. (2015). Benefits of dams, http://www.fema.gov/benefits-dams. Cited 26 Jan 2016.

Whitham, G. (1974). *Linear and nonlinear waves*.

Wubs, F. (1988). *Numerical solution of the shallow-water equations*.

Xing, Y. and Shu, C. (2005). High order finite difference WENO schemes with the exact conservation property for the shallow water equations, *J. Comput. Phys.* **208**, 1, pp. 206–227.

Yilmaz, O., Korobkin, A., and Iafrati, A. (2013). The initial stage of dam-break flow of two immiscible fluids. Linear analysis of global flow, *Appl. Ocean Res.* **42**, pp. 60–69.

Zhou, J., Causon, D., Ingram, D., and Mingham, C. (2002). Numerical solutions of the shallow water equations with discontinuous bed topography, *Int. J. Numer. Meth. Fluid.* **38**, 8, pp. 769–788.

Zoppou, C. and Roberts, S. (2000). Numerical solution of the two-dimensional unsteady dam break, *Appl. Math. Model.* **24**, 7, pp. 457–475.

Index

155

Printed in the United States
by Baker & Taylor Publisher Services